Pharmaceutics-I

(Dispensing Pharmacy)

Practical Notebook

(For B.Pharm Students)

By

A.K. Gupta M.Pharm (Pharmaceutics)
Ex-Head, Department of Pharmacy,
Govt. Polytechnic for Women, Chandigarh

and

V.K. Jain

Name of Student ....

University Roll No. ....

College ....

....

CBS

CBS Publishers & Distributors Pvt. Ltd.

New Delhi • Bengaluru • Chennai • Kochi • Kolkata • Mumbai
Hyderabad • Nagpur • Patna • Pune • Vijayawada

ISBN: 81-239-1455-5

First Edition: 2007
Reprint: 2011, 2016

Published by:
Satish Kumar Jain for CBS Publishers & Distributors Pvt. Ltd.,
4819/XI Prahlad Street, 24 Ansari Road, Daryaganj, New Delhi - 110002
delhi@cbspd.com, cbspubs@airtelmail.in • www.cbspd.com
Ph.: 23289259, 23266861, 23266867 • Fax: 011-23243014

Corporate Office: 204 FIE, Industrial Area, Patparganj, Delhi - 110 092
Ph: 49344934 • Fax: 011-49344935
E-mail: publishing@cbspd.com • publicity@cbspd.com

Branches:
- ***Bengaluru:*** 2975, 17th Cross, K.R. Road, Bansankari 2nd Stage, Bengaluru - 70 • Ph: +91-80-26771678/79 • Fax: +91-80-26771680 E-mail: cbsbng@gmail.com, bangalore@cbspd.com
- ***Chennai:*** No. 7, Subbaraya Street, Shenoy Nagar, Chennai - 600030 Ph: +91-44-26681266, 26680620 • Fax: +91-44-42032115 E-mail: chennai@cbspd.com
- ***Kochi:*** Ashana House, 39/1904, A.M. Thomas Road, Valanjambalam, Ernakulum, Kochi • Ph: +91-484-4059061-65 Fax: +91-484-4059065 • E-mail: cochin@cbspd.com
- ***Kolkata:*** 6-B, Ground Floor, Rameshwar Shaw Road, Kolkata - 700014 Ph: +91-33-22891126/7/8 • E-mail: kolkata@cbspd.com
- ***Mumbai:*** 83-C, Dr. E. Moses Road, Worli, Mumbai - 400018 Ph: +91-9833017933, 022-24902340/41 • E-mail: mumbai@cbspd.com

Representatives:
- Hyderabad: 0-9885175004
- Nagpur: 0-9021734563
- Patna: 0-9334159340
- Pune: 0-9623451994
- Vijayawada: 0-9000660880

Printed at:
J.S. Offset Printers, Delhi

Foreword

Pharmacy Council of India has implemented New Education Regulation 1991 w.e.f. 1993. Pharmaceutics-I is one of the most important subjects in E.R. 1991 which requires that the pharmacist should be able to dispense all types of formulations used by patients. The manual written by Shri A.K. Gupta contains the theory and method of dispensing of different types of formulations as per syllabus. Efforts have been made by the author to write in simple and precise manner to make the subject understandable by the students of B.Pharm standard. This manual will be equally useful to the pharmacy students, teachers and working pharmacists. I recommend this manual to B.Pharm 1st year students and suggest that they should dispense the maximum number of prescriptions themselves.

S.S. Bajaj

Preface

It gives me an immense pleasure that the first edition of Pharmaceutics-I Practical Manual was warmly accepted by the teachers and students of Pharmacy as it made their work quite easy and they could devote much of their time actually to the practical work. I am highly thankful to all of them for their appreciation and suggestions sent from time to time.

With the introduction of Education Regulation of Pharmacy Council of India, it was necessitated to revise and bring this practical manual upto the expectations of teachers and students of Pharmacy.

As per syllabus, the prescriptions covering various topics like powders, capsules, mixtures, emulsions, ointments, E.N.T. preparations, liniments, lotions, incompatibilities, suppositories and cosmetics are covered. Along with the exercises, examples with their brand names of marketed preparations are given which will make the subject interesting and impart knowledge of the marketed preparations.

I am sure this practical manual will prove useful to the teachers, students and working pharmacists. It will be helpful to the students as after finishing their practical work they will directly write the calculations, method of preparation in their own language and also make a beautiful label on the blank left hand side page of this manual instead of writing it in a separate notebook thus saving their valuable time and money. By using this printed practical manual they will be able to devote much of their time for the actual practical work. Here theoretical part concerning every topic is also given in the starting of each and every individual topic. The students will be able to prepare for viva voce by consulting the practical manual enabling them to secure good marks in Viva Voce.

Thanks are due to all the teachers, students and working pharmacists throughout the country who had given me encouragement, appreciation and suggestion and inspired me to write the new edition of this manual.

THE PRESCRIPTIONS GIVEN IN THE MANUAL ARE FOR PRACTICAL PURPOSE ONLY.

I am highly thankful to Shri S.S. Bajaj, Ex-Head, Department of Pharmacy, Govt. Polytechnic for Women, Chandigarh for writing the foreword, for the encouragement and sincere guidance he has been giving from time to time.

My thanks are due to Shri Satish Kumar Jain and Shri Vinod Kumar Jain and the whole team of M/s CBS Publishers & Distributors for their keen interest in publishing this manual in a short span of time.

Valuable suggestions and criticism from teachers, students and working pharmacists for further improvement in the quality of this manual will be thankfully received and acknowledged.

2007

Ashok K. Gupta

Contents

Index

S.No.	Name of the Experiment	Date	Page	Remarks

Index

S.No.	Name of the Experiment	Date	Page	Remarks

Index

S.No.	Name of the Experiment	Date	Page	Remarks

Index

S.No.	Name of the Experiment	Date	Page	Remarks

1

Instructions

1. Always wear a good quality, neat and clean white apron whenever you are to do work in the laboratory. This will give good impression to the fellow beings and you will potect your clothes from spoilage.
2. Always bring practical note book neat and tidy to the laboratory.
3. Before starting the prescription work, thoroughly understand the ingredients, method of preparation and apparatus required. In case of doubt must consult your teacher. No guess work should be done which may lead to serious results.
4. The students will be provided with the following apparatus for which he will be held responsible for any loss, breakage or missing articles. Therefore, he must keep the apparatus under lock in the cupboards provided on each seat. He must clean the apparatus after use and then store in the cupboard for future use :
    1. Dispensing balance
    2. Weight box
    3. Dispensing bottles
    4. Pestle and mortar (wedge wood)
    5. Pestle and mortar (glass)
    6. Ointment slab
    7. Ointment spatula
    8. Horn/bone spatula
    9. Beakers 250 ml and 100 ml
    10. Measuring cylinders 5 ml, 10 ml and 100 ml
    11. Porcelain dish
    12. Glass funnel
    13. Glass rods
    14. Ointment jar/jelly pot
    15. Wide mouth bottles
    16. Tripot stand
    17. Wire gauge
    18. Water bath
    19. Sand bath
    20. Suppository mould
    21. Suppository bath and cone
    22. Dusters
5. Thoroughly clean the apparatus required in the practical work, uncleaned apparatus may not give good results. Only specified apparatus required for particular work should be used.

6. Keep the balance in front of you, clean the pans, adjust the instrument if necessary, place a piece of butter paper on either side of the pans. Powdered chemicals and liquids should be weighed on a watch glass or in a weighing bottle and not directly on the pans to avoid spoilage of the pans.

   Always use forceps for lifting the weights. Put weights on the right hand pan and the material on the left hand pan.
7. Do not weigh less than 100 mg on the dispensing balance because these balances are not very sensitive. Smallar amounts less than 100 mg should be weighed on a high quality analytical balance or firstly triturations prepared and then weighed.
8. Liquids must be measured accurately with the smallest possible measure. The whole of the quantity of a liquid must be measured collectively and the volume should not be splitted into two measurements which may increase errors.
9. Collect the materials required to be weighed on the left hand side of the balance. Arrange them in the sequence in which they are to be weighed. Read the label of the container three times (i) when removing it from the shelf, (ii) when taking out the material for weighing, (iii) when placing the contaier back on to the shelf.

   While taking out the material from the bottle, hold the stopper or the cork in the left hand and the bottle in the right hand. Do not place the cork on the table. As soon as the material has been removed, immediately replace the cork.
10. All the dosage forms should be dispensed in suitable containers specially meant for the purpose. Container should be almost of the same size as that of medicament to be dispensed.
11. After filling the container, it should be corked, labelled and polished. Good quantity of paper and adhesive should be used for fixing the label on the container.
12. All the entries must be recorded as soon as the practical work is completed and practical note book complete in every respect should be submitted on the same day to the teacher concerned for marking and signature.
13. Never throw waste papers and burnt match sticks in the sink, this will block the sink which may lead to overflowing. Put these waste materials in the dust bins.
14. After the work is finished the students should clean the apparatus and store in the cupboard. The seats should be cleaned with duster and see that the seats are kept clean during practical work as well.
15. The water taps and gas taps should be tightly closed after the work is over. Electrical switches should be put off before leaving the laboratory.

Important note

Students are advised to write date, exercise number and calculated quantities for required amount of preparation to be prepared on the left hand side of the note book; translate the latin language; write procedure in the past tense and make a beautiful label (uniform in size etc. throughout the practical note book) at the end of each exercise.

2

Prescription

Prescription is an order written by a physician, dentist or any other registered medical practitioner to a pharmacist to compound and dispense a specific medication for the patient. The order is accompanied by directions for the pharmacist that what type of preparation is to be prepared and how much is to be is prepared. It is also accompanied with the directions for the patient that how much medicament is to be taken, how many times or at what time and how it is to be taken.

The prescription provides a common link of mutual interest between the physicians, the pharmacist and the patient. It is the duty of the pharmacist to serve the medication needs of the patient according to the intention of the prescriber. It is not sufficient that the pharmacist should only compound the specific medication but he should make the patient understand about the proper administration of the drug and ensure that the patient sticks to these instructions. At the same time the pharmacist must maintain and respect the confidentiality of both the physician regarding the treatment given as well as that of the patient regarding the nature of his illness and the medication taken by him.

Prescriptions were generally written in Latin language which was understood all over the world, so that the ingredients of the prescriptions remain unknown to the patient to avoid self-medication. Still the use of Latin abbreviations in the prescription writing is very common specially in dosage instructions.

The modern tendency in prescribing drugs is to make prescriptions simple. Instead of prescribing a mixture of numerous drugs, the prescription of a single drug with definite and specific action is desired. More stress is laid on the readymade preparations which are available in the market.

Parts of a prescription

A typical prescription consists of the following parts :

1. Date

Date must be written on the prescription by the prescriber at the same time when it is written. This helps the pharmacist to know that when the medicines were dispensed last time if the prescription is brought for redispensing of the medicines.

2. Name, age, sex and address of the patient

These particulars help the pharmacist that the medicine has been delivered to the right person and correct dose has been given to the children. The address of the patient is recorded to help for any reference at a later stage or to contact the patient in case of any emergency.

3. Superscription

It is represented by a symbol ℞ an abbreviation of Latin word 'recipe' which means 'take thou' or 'you take'. The line on ℞ is said to designate Jupiter, the God of healing.

4. Inscription

This is the main part of prescription. It contains the names and quantities of the prescribed ingredients. It is usually divided into :

(a) The base or the active medicament which is intended to produce the therapeutic effect.
(b) The adjuvant which is included either to enhance the action of the medicament or to make the product more palatable.
(c) The vehicle which is either used to dissolve the solid substance and/or to increase the volume of the preparation for ease of administration.

5. Subscription

This part contains the prescriber's directions to the pharmacist regarding the dosage form to be prepared and number of doses to be dispensed.

6. Signature

It is usually abbreviated as "Sig" and consists of the directions to be given to the patient regarding the method of administration or application, quantity to be used, number of times in a day, at what time and the vehicle with which it is to be used. These directions must be transferred to the label of the bottle.

7. Signature of the prescriber

The prescription must be signed by the prescriber with his own hand. His address and registration number should be written in the case of dangerous drugs.

An example of a typical prescription

General Hospital

Date : 17.7.1990
Name : Shri Parmod Kumar
Age : 45 years Sex : Male
Address : 927, Sector 23, Chandigarh

℞

Sodium bicarbonate	1.0 gm
Aromatic spirit of ammonia	1.5 ml
Cardamom tincture	1.0 ml
Spirit chloroform	1.5
Water to	15.0 ml

Fiat : Mistura. Mitte doses tres.
Sig: Unus post cibos sumenda

M.D. Goyal
Regd. No. 15557

Handling of prescription

1. Receiving

The prescription should be received by the pharmacist and read carefully. If there is any doubt regarding prescription, he must consult the prescriber. There should not be guess work regarding spellings etc. because this may lead to serious consequences. Therefore, doubts must be cleared before compounding any prescription.

2. Collecting the materials

Read the label on the bottle of the ingredient at the time of removing from the shelf, taking out the ingredients from the container and again at the time of replacing the container back on the shelf to avoid any mistake.

3. Compounding

Compounding is the most important phase in handling the prescription. In this case proper drug is dispensed in a suitable form. This can be achieved only if accuracy, cleanliness and if proper techniues are observed in the preparation of any medication. Only one prescription should be compounded at one time.

The finished product should be transferred to a suitable container depending on the nature and quantity of the medicament to be dispensed and the method of its use. Various types of containers used for different types of formulations are described below :

(i) Round vials : for tablets and capsules.
(ii) Oval prescription bottles : for liquids of low viscosity e.g. mixtures, emulsions etc.
(iii) Wide mouth bottles : for filling liquids of high viscosity, large quantities of tablets, capsules and bulk powders.
(iv) Coloured flutted bottles : for preparations meant for external use e.g. liniments and lotions.
(v) Ointment jars and collapsible tubes : for ointments, creams or any other semi-solid dosage form.
(vi) Paper wrapper and envelops : for oral powders dispensed as individual doses.
(vii) Dropper bottles : for eye drops, ear drops or other liquids to be administered by drops.
(viii) Sifter top containers : for dusting powders.

The container should be selected approximately of the same volume as that of medication to be dispersed.

4. Labelling

The filled containers should be suitably labelled. A good quality of paper and adhesive should be used for labelling the containers. The size of the label should be proportional to the size of the container and should be hand written in capital letters or typed. The label should be affixed on the smooth surface (in the case of dispensing bottles) and in the centre of the bottle leaving sufficient margin towards the top and bottom of the bottle. The following information should be written on all the labels:

(i) Type of the preparation : The mixture, the emulsion, the powder etc. Its quantity should also be mentioned.

(ii) For : Name of the patient, age and sex.
(iii) Registration no.
(iv) Date of dispensing.
(v) Expiry date if any.
(vi) Direction for its use.
(vii) Storage conditions.
(viii) Name and address of the pharmacy.
(ix) In case of liquid preparations which require previous shaking e.g. emulsions, suspensions, liniments, lotions etc. must be attached with secondary label "shake the bottle before use".
(x) The prescriptions meant for external use e.g. liniments, lotions, ointments, eye drops, ear drops etc. must be attached with secondary label "for external use only".

The sample labels are given below :

THE POWDER
(5 × 250 mg)
For : Mrs. X.Y.Z. Age : 25 yrs. (F)
Regd No. 47
Date of dispensing :
Directions : One to be taken when the pain is severe
Dispensed by :
Name and address of pharmacy

THE MIXTURE
(90 ml)
For : Mr. X.Y.Z. Age : 40 yrs. (M)
Regd No. 247 **Shake the bottle before use**
Date of dispensing :
Directions : One Tablespoonful to be taken three times a day.
Dispensed by :
Name and address of pharmacy

THE OINTMENT
(40 gm)
For : Miss A.B.C. Age : 20 yrs. (F)
Regd No. 307 : **For external use only**
Date of dispensing :
Directions : To be applied to the skin as directed.
Dispensed by :
Name and address of pharmacy

THE LINIMENT
(60 ml)
For : Mr. X.Y.Z. Age : 28 yrs. (M)
Regd No. 327 **For external use only**
Date of dispensing :
Directions : To be applied to the affected part with rubbing
Dispensed by : **Shake the bottle before use**
Name and address of pharmacy

5. Marking the dose

A dose marking should be applied to all the multidose containers of liquid preparations e.g. mixtures and emulsions for the convenience of the patient while taking a dose. After filling the container a piece of paper is cut upto the level of the liquid from the centre of the curvature of the bottle. This paper is then divided into equal parts according to the number of doses to be administered which should be given a deep cut to clearly indicate the level of the dose. This paper should be fixed to the bottle.

6. Finishing

After filling and labelling, the container should be corked, thoroughly polished to remove the finger prints and then dispensed.

3

Latin Terms

Latin Terms and Abbreviations Commonly Used in Prescription Writing

Latin term or phrase	*Abbreviation*	*English meaning*
Ad	ad	to, up to
Ad libitum	ad, lib	at pleasure, as desired
Admove	admove	apply
Agita	agit	shake, stir
Alter	alt	the other, alternate
Alternis horis	alt. hrs.	alternate hours
Ana	aa.	of each
Ante	a.	before
Ante cibos	a.c.	before meals
Applicandus	applicand	to be applied
Aqua	aq.	water
Aqua bulliens	aq. bull.	boiling water
Aqua destillata	aq. dest.	distilled water
Auris dextra	a.d.	right ear
Auris laeva	a.l.	left ear
Bis in die	b.i.d.	twice a day
Capsula	caps	capsule
Capiendus	capiend	to be taken
Cataplasma	cataplasm	poultice
Charta	chart	powder paper
Cibos	cibos	food, meals
Cochleare amplum Cochleare magnum Cochleare maximum	coch amp coch mag coch max	one table spoonful
Cochleare medium Cochleare modicum	coch med coch mod	one desert spoonful
Cochlear minimum Cochlear parvum	coch min coch parv	one teaspoonful
Collunarium	collunar	a nose wash
Collutorium	collut	a mouth wash

(Contd.)

Latin term or phrase	*Abbreviation*	*English meaning*
Collyrium	collyr	an eye wash
Congius	cong	a gallon
Cum	c	with
Cum duplo	c. dup.	with twice as much
Cum parte aequale	c. pt. aeq.	with an equal quantity
Cyanthus	cyanth	a glass
Dexter	dext	right
Divide	div.	divide
Dolore urgente	dol. urg.	when the pain is severe
Emulsio	emul	an emulsion
E	–	with
Ex	–	out of
E. lacte	e. lact	with milk
Ex. aqua	ex. aq.	with water
Ex. modo prescripto	e.m.p.	in the manner prescribed
Fiat, fiant	ft.	make, let it be made
Granum, grana	gr.	a grain
Gutta, guttae	gtt.	a drop, drops
Hora	h.	an hour
Hora somni	h.s.	at bed time
In dies	In.d.	daily
Inter cibos	i.c.	during meals
Injectio	Inj.	an injection
Laevo	l.	left
Levis	lev.	light
Linimentum	lin.	a liniment
Liquor	liq.	solution
Mane	m.	morning
Minimum	min.	a minim
Misce	m.	mix, let (it) be mixed
Mistura	mist	a mixture
Mitte	mitt	send
Mitte tales	mitt tal	send such
Modo dicto	m. dict	as directed, as stated
Modo prescripto	m. pres	as prescribed
More dicto	m. dict	as directed
Nebula	nebul	a spray
Nocte maneque	noct. maneq	night and morning
Non repetatur	non rep, n.r.	do not repeat
Octarius	o.	a pint
Oculo utro	o.u.	each eye
Oculus dexter	o.d.	right eye

(Contd.)

Latin term or phrase	*Abbreviation*	*English meaning*
Oculus laevus	o.l.	left eye
Oculus sinister	o.s.	left eye
Omni	omn.	every
Omni hora	omn. hor, o.h.	every hour
Omni quadranta hora	omn. quadr. hor	every quarter of an hour
Omni quarta hora	omn, 4 hrs.	every four hours
Omni secunda hora	omn. 2 hrs.	every two hours
Omni mane	o.m.	every morning
Os, oris	o.s.	mouth
Per os.	p.o.	orally, by mouth
Pot. cibos	p.c.	after meals
Pulvis	pulv	powder
Quantum sufficiat	q.s.	As much as sufficient
Quarter in die	q.i.d.	four times a day
Recipe	Rx	take
Secundum artum	s.a.	according to the art
Semis, Semi	ss.	half
Signa, signetur	sig.	write
Si opus sit	s.o.s.	when necessary
Solve	–	dissolve
Solutio	sol	a solution
Statim	stat	immediately
Sumendus	sum	to be taken
Suppositorium	supp.	a suppository
Tabella, tabletta	tab	a tablet
Talis, tales	tal	such
Ter in die	t.i.d.	three times a day
Uncia	–	an ounce
Unguentum	ung	an ointment
Utendus	u. or utend	to be used
Unus	i	one
Du	ii	two
Tres	iii	three
Quatuor	iv	four
Quinque	v	five
Sex	vi	six
Septem	vii	seven
Octo	viii	eight
Novem	ix	nine
Decem	x	ten
Undecim	xi	eleven
Duodecim	xii	twelve
Quindecim	xv	fifteen

4

Weights and Measures

There are two systems of weights and measures (a) the imperial system, (b) the metric system, with which the pharmacist must be familiar. The imperial system is an old system based on arbitrary and unrelated units e.g. grains, drachms, ounces and gallons whereas the metric system or decimal system is based on related and rationally derived units e.g. milligrams, grams, centimeters, meters, millimetres, litres etc. Because of its easier calculations, greater accuracy and flexibility and use in other sciences, now a days this is the most widely used system by official agencies.

At present still a large number of physicians trained to use the imperial system prescribe the drugs in the old system and some hospitals still retain it as the local standard. Some drugs are prescribed in fractional doses (1/200, 1/150, 1/100 gr.). The bottles for liquids are still manufactured to contain ounce measurements rather than milliliters.

Due to the above mentioned reasons, it is still necessary to be familiar with both the systems which are described in detail as follows :

IMPERIAL SYSTEM

Imperial system is divided into two systems.

(i) Avoirdupois system.

(ii) Apothecaries system.

Avoirdupois system

According to this system the standard unit for weighing is pound and all other measures of mass are derived from pound. It is represented by lb.

1 lb = 16 oz (avoir)

1 lb = 7000 grains

1 oz = 7000/16 = 437.5 grains

Apothecaries system

It is also known as troy system. The standard weight in this system is grain.

20 grain = 1 scruple

60 grain = 1 drachum

480 grain = 1 ounce (Apothe)

8 drachum = 1 ounce (Apothe)

12 ounces (Apothe) = 1 pound (Apothe)

5760 grain = 1 pound (Apothe)

Abbreviations commonly used in weighing

Latin name	*Symbol*	*English name*	*Equal to*
Granum	gr	grain	1 grain
Scrupulus	∉	Scruple	20 grain
Drachma		drachm	60 grain
Uncia	oz	ounce (Avoir)	437.5 grain
Uncia		ounce (Apothe)	480 grains
Libra	lb	pound (Avoir)	7000 grains
Libra	lb	pound (Apothe)	5760 grains

Measures of capacity

Standard units for capacity are same in avoirdupois as well as apothecaries system. The standard unit is gallon and all other measures of capacity are derived from gallon.

1 gallon	= 160 fluid ounces	
1/4th of a gallon	= 1 quart	= 40 fl. ounce
1/8th of a gallon	= 1 pint	= 20 fl. ounce
1/160th of a gallon	= 1 fl. ounce	
1/8th of one fl. ounce	= 1 fl. drachm	
1/60th of one fl. drachm	= 1 minim.	
1 fluid ounce	= 480 minim.	
1 fluid drachm	= 480/8	= 60 minim.

Abbreviation commonly used in measures of capacity

Latin name	*Symbol*	*English name*	*Equal to*
Minimum	m	minim	1 minim
Fluidrachma		fl. drachm	60 minim
Fluidunicia		fl. ounce	480 minim
Octarius	O	pint	20 fl. ounces
Congius	C	gallon	160 fl. ounces

METRIC SYSTEM

Standard unit of measures of mass (weight) is kilogram and all other measures of mass are derived from kilogram

1 Kilogram (kg)	= 100 gm	
1 Hectogram (hg)	= 100 gm	
1 Decagram (dag)	= 10 gm	
1 Gram (gm)	= 1 gm	
1 Decigram (dg)	= 0.1 gm	= 100 mg
1 Centigram (cg)	= 0.01 gm	= 10 mg
1 Milligram (mg)	= 0.001 gm	= 1 mg
1 Microgram (mg, mcg)	= 1/1000 mg	

Measures of capacity

Standard unit for measures of capacity (volume) is litre and all other measures of capacity are derived from litre.

1 litre (lt)	= 1000 millilitre (ml)

Domestic measures

1 drop	= 1 minim	= 0.06 ml
1 tea spoonful	= 1 fl. drachm	= 4 ml
1 desert spoonful	= 2 fl. drachm	= 8 ml
1 table spoonful	= 4 fl. drachm	= 15 ml
2 table spoonful	= 1 fl. ounce	= 30 ml
1 wine grassful	= 2 fl. ounce	= 60 ml
1 tea cupful	= 4 fl. ounce	= 120 ml
1 tumblerful	= 8 fl. ounce	= 240 ml

Conversion factors

1 grain	= 64.8 mg	= 65 mg (for all practical purposes)
1 drop	= 1 minim	= 0.06 ml (for all practical purposes)
1 fl. ounce	= 29.57 ml	= 30 ml (for all practical purposes)
1 gram	= 15.4 gr	= 15 gr (for all practical purposes)
1 milligram	= 1/65 gr	= 1/65 gr (for all practical purposes)
1 litre	= 33.8 fl. ounce	
1 kilogram	= 2.2 pound	

5

Powders

Powders are the solid dosage form of medicament in which the drug or drugs are dispensed in a finely divided state. They are available in crystalline or amorphous form. Whenever crystalline substances are to be dispensed in powder form, they must be reduced to a fine powder before mixing them with other substances. Powders may be dispensed as divided powders (single doses) by wrapping individually in suitable size papers or as bulk powders in suitable containers.

Classification

Powers may be classified as follows :

1. Simple powders
2. Compound powders
3. Powders enclosed in capsules
4. Compressed powders (Tablets)
5. Granular effervescent powders
6. Powders for external use

1. Simple powders

Simple powders are those powders when only one ingredient is present. They are generally wrapped in individual papers and dispensed.

2. Compound powders

In compound powders more than one ingredients are mixed together in the ascending order of their weights. Whenever crystalline substances are to be incorporated they must be reduced to a fine powder, mixed and divided into required number of powders. They are dispensed as individual doses by wrapping the powder in suitable size paper or it may be dispensed as bulk powder in a suitable sized screw capped wide mouthed bottle with specific instructions for its use. For hygroscopic powders double wrapping should be done in which inner wrapper should be waxed paper.

Since the wastage of powders during weighing, mixing and dividing is unavoidable so calculate for one extra powder than required. If this produces quantities in which fractions are involved, calculate for sufficient number of extra powders to produce directly weighable quantities.

Since the dispensing balances are not very sensitive so quantities less than 100 mg cannot be weighed accurately on such balances, therefore when very small quantities of drugs are to be used they must be first diluted with an inert substance to increase the quantity of the drug so as to make it a weighable quantity. Potent substances like Codeine phosphate, Arsenic trioxide, Hyoscine hydrobromide etc. are dispensed in this way.

Potent substances

Substances having a minimum dose of less than 60 mg should be regarded as potent substances. These should be weighed either on a chemical balance or upon a delicate pair of dispensing balance specially reserved for the purpose. In no case quantities less than 60 mg be weighed on such balance. When quantities less than 60 mg are required, triturations should be prepared and then weighed.

Bulk powders

When large quantities of powders are dispensed in a container and the patient is asked to use a measured quantity of the drug as a dose. Generally these powders are dispensed in wide mouth containers along with a suitable measure for easy removal of the contents.

Effervescent granules

These are the powders made in the form of granules which when added to water produce effervescence with the evolution of carbon dioxide. Generally these powders are used as antacids.

Effervescent granules usually contain Sodium bicarbonate, Citric acid and Tartaric acid. They may also contain certain medicament and sweetening agent like sucrose or saccharin.

Powders for external use

They are generally known as dusting powders and are meant for external application to the skin for their antiseptic, antiperspirant lubricant, protective, absorbent and astringent purposes. The dusting powders must be passed through a very fine sieve to remove gritty particles. They may be packed in sifter top containers or wide mouthed containers.

Exercise No. 1

For :
Age :
Address :

℞

Aspirin	300 mg
Paracetamol	150 mg

Fiat pulvis. Mitte tales quarta.

Sig : Unus more dicto sumenda.
Type : Oral divided powder.

Theory

Since there will be little wastage during weighing, mixing and dividing the powders, so calculate for one extra powder.

Procedure

Weigh the calculated amount of Aspirin and paracetamol. Mix them thoroughly. Weigh 450 mg of the mixed powder for each powder. Prepare such four powders and discard the rest. Wrap in suitable size powder papers, label and dispense.

Use

It is used as an analgesic, antipyretic and anti-inflammatory preparation.

Aspirin : Analgesic, antipyretic and anti-inflammatory.
Paracetamol : Analgesic and antipyretic.

Exercise No. 2

For :
Age :
Address :

℞

Aspirin	300 mg
Paracetamol	150 mg
Caffeine	50 mg

Fiat pulvis. Mitte tales octo.

Sig : Unus dolore urgente sumenda
Type : Oral divided powder.

Theory

Since there will be little wastage during weighing, mixing and dividing so taking into consideration this wastage, calculate for one extra powder.

Procedure

Weigh the calculated quantities of Aspirin, Paracetamol and Caffeine. Mix, in the ascending order of their weights. Weigh out 500 mg of the mixed powders for each powder. Prepare such eight powders and discard the rest. Wrap each dose in individual powder paper, label and dispense.

Uses

It is used as an analgesic, antipyretic and anti-inflammatory preparation.

Aspirin : Analgesic, antipyretic and anti-inflammatory.
Paracetamol : Analgesic and antipyretic.
Caffeine : C.N.S. stimulant. It is some way potentiates the actions of other ingredients.

Exercise No. 3

For :
Age :
Address :

℞

Calcium carbonate	37.5 gm
Heavy magnesium carbonate	37.5 gm
Sodium bicarbonate	12.5 gm
Bismuth carbonate	12.5 gm

Make a powder.

Label : The Compound Bismuth Powder. Send 50 gm.
Sig : 5 gm bis in die sumenda.
Type : Bulk powder.

Procedure

Weigh accurately all the ingredients and mix them in the ascending order of their weights. Pack the mixed powder in wide mouthed container, label and dispense.

Uses

This preparation is used as an antacid.

Marketed preparation by :
Geoffrey Manners & Co. Ltd., Mumbai - 400038
Bismag powder
Contains :
Sodium bicarbonate, heavy magnesium carbonate, light magnesium carbonate, calcium carbonate.

Exercise No. 4

For :
Age :
Address :

℞

Sodium bicarbonate	25.0 gm
Heavy magnesium carbonate	25.0 gm
Magnesium trisilicate	25.0 gm
Chalk in powder	25.0 gm

Make a powder.

Label : The Compound Magnesium Trisilicate Oral powder. Send 25 gm.
Type : Compound oral bulk powder.

Procedure

Weigh accurately Sodium bicarbonate, Heavy magnesium carbonate, Magnesium trisilicate and Chalk in powder. Mix them thoroughly in the ascending order of their weights. Pack in wide mouthed container, label and dispense.

Uses

This preparation is used as an antacid.

Marketed preparation by :
Parke-Davis (India) Ltd., Mumbai-400025.
Takazyme powder
Each contains :

Magnesium carbonate light	16%
Magnesium trisilicate	8%

(Aspergillus Oryzae Enzyme equivalent to 1.07% Alpha Amylase)

Exercise No. 5

For :
Age :
Address :

℞

Sodium chloride in powder	0.500 gm
Potassium chloride in powder	0.750 gm
Sodium bicarbonate	0.750 gm
Dextrose	20.000 gm

Make a powder.

Label : The Compound Sodium Chloride and Dextrose Powder.
Sig : More dicto sumenda.
Type : Compound oral powder.

Procedure

Powder all the ingredients if already not in fine powder. Weigh each ingredient accurately and mix them thoroughly in the ascending order of their weights. Wrap in powder paper in which inner wrapper should be waxed paper or dispense in individual well closed containers intended for use only once.

Uses

This powder is used as an electrolyte powder as an oral rehydration therapy i.e. during dehydration. It is to be used by dissolving the contents in the prescribed amount of water and solution used at intervals as described by the physician.

Marketed preparations by :

1. Parke Davis (India) Ltd., Mumbai - 400025
   Prequest Forte Sachet
   Each contains :

Sodium chloride	11.700%
Potassium chloride	4.967%
Sodium bicarbonate	8.400%
Dextrose anhydrous	70.636%

2. Rays Laboratories Pvt. Ltd., Kolkata - 700007
   Raylyte powder
   Each 27.5 gm contains :

Sodium chloride	3.5 gm
Sodium bicarbonate	2.5 gm
Potassium chloride	1.5 gm
Dextrose	20.0 gm

Exercise No. 6

For :
Age :
Address :

℞

Rhubarb, in powder	25.0 gm
Ginger, in powder	10.0 gm
Light magnesium carbonate	32.5 gm
Heavy magnesium carbonate	32.5 gm

Make powder.

Label : The Gregory's powder (Compound Rhubarb powder) send 25 gm.
Directions : 0.5 to 5.0 gm to be taken twice in a day.
Type : Bulk oral powder.

Theory

Since it is difficult to finely powder the vegetable drugs in a pestle and mortar on laboratory scale, therefore such powders are used which are already available in the powder form. Light magnesium carbonate is used because it disperses easily in the liquids. Heavy magnesium carbonate is used to avoid bulkiness of the preparation. Rhubarb powder acts as a laxative. Ginger powder imparts flavouring properties and prevent gripping action produced by rhubarb.

Procedure

Weigh the calculated amount of rhubarb, in powder, ginger, in powder, light magnesium carbonate and heavy magnesium carbonate. Mix these powders in the ascending order of their weights. Transfer the mixed powders in suitable air tight containers, label and dispense.

Uses

This powder is used as laxative.

Exercise No. 7

For :
Age :
Address :

℞

Arsenic trioxide 0.002 gm

Fiat pulvis. Mitte tales novem.

Sig : Unus more dicto sumenda.
Type : Potent oral divided powder.

Theory

Since there will be little wastage of powders during weighing, mixing and dividing therefore calculate for one extra powder than required.

The quantity of Arsenic trioxide prescribed is very very small (potent substance) and it is not possible to weigh such a small quantity on dispensing scale therefore 100 mg (minimum amount) of Arsenic trioxide will have to be weighed on the dispensing balance. For dilutions lactose is used as diluent because it is white, sweet in taste, easily available, cheap and compatible with majority of the drugs.

Procedure

Weigh 100 mg of Arsenic trioxide and mix it with 400 mg of lactose = 500 mg total by means of trituration i.e. add about 100 mg of lactose to 100 mg of Arsenic trioxide, mix them intimately, then gradually add the remaining amount of lactose with intimate mixing after each addition. If both the substances will be mixed with each other all atonce, there will not be uniform dispersion of one substance in the other. This total 500 mg of the triturate will contain 100 mg of Arsenic trioxide. Weigh out 100 mg of this triturate which will contain 20 mg of Arsenic trioxide. To this 100 mg of triturate add 900 mg lactose = 1 gm total by means of trituration. Out of this triturate weigh 100 mg for each powder which will contain 0.002 gm (2 mg) of Arsenic trioxide. Wrap in powder papers as individual doses, label and dispense.

Uses

Arsenic trioxide is used as bitter tonic.

Exercise No. 8

For :
Age :
Address :

℞

Prednisone 8 mg

Make a powder.
Send such four powders.

Signa : One to be taken as directed.
Type : Simple divided powder.

Theory

Since there will be little wastage during weighing, mixing and dividing therefore calculate for one extra powder.

The quantity of Prednisone prescribed is very small (potent substance) and it is not possible to weigh such a small quantity on dispensing balance, therefore 100 mg (minimum amount) of Prednisone will have to be weighed which will be diluted with lactose by means of trituration.

Procedure

Weigh 100 mg of Prednisone and mix it with 900 mg of lactose by means of trituration = 1 gm total. Take out 400 mg of this triturate which will contain 40 mg of Prednisone. To this 400 mg of triturate add 100 mg lactose and mix thoroughly. Weigh out 100 mg for each powder. Prepare such 4 powders and discard the rest. Pack in powder papers, label and dispense.

Uses

It is used as corticosteroid.

Exercise No. 9

For :
Age :
Address :

℞

Phenobarbitone sodium 15 mg

Send such 4 powders.

Sig : Unus omni nocte sumenda.
Type : Simple oral divided powder.

Theory

Since there will be little wastage of powder during weighing, mixing and dividing the powder therefore calculate for one extra powder than required.

The quantity of Phenobarbitone sodium prescribed is very small and it is not possible to weigh such a small quantity on dispensing scale therefore 100 mg (minimum amount) of Phenobarbitone sodium will have to be weighed which will be diluted with lactose by means of trituration.

Procedure

Weigh 100 mg of Phenobarbitone sodium and mix it with 900 mg of lactose by means of trituration = 1 gm total. Weigh out 150 mg of this mixture which will contain 15 mg of Phenobarbitone sodium. Prepare such four powders each weighing 150 mg of the triturate, discard the rest, wrap in powder papers, label and dispense.

Uses

Phenobarbitone sodium is used as hypnotic, sedative and anticonvulsant.

Exercise No. 10

For :
Age :
Address :

℞

Menthol	5 gm
Camphor	5 gm
Ammonium chloride	30 gm
Light Magnesium carbonate	60 gm
Fiat Insufflatio. Mitte	50 gm

Sig : Pro naso
Type : Compound powder

Theory

This prescription contains liquefiable substances which when combined together form a liquid due to the formation of a compound having lower melting point than the individual substances. The substances which behaves in this way are known as eutectic substances and the mixture so formed after combining the substances is known as eutectic mixture. The liquefiable substances include menthol, thymol, camphor, phenol, antipyrine, acetophenatidine, aspirin, salol and acetanilide. When any two of them are mixed together they liquefy. To deal with such substances they should be powdered and mixed separately with a portion of suitable diluent and then gently mixed together. When in addition to liquefiable substances, other ingredients are also present, the liquefiable substances should first be mixed together to form a eutectic mixture, then to this an absorbent is added and other ingredients in powder form are mixed. The most commonly used absorbents include light magnesium oxide but starch, talc, lactose, calcium phosphate etc. may also serve the purpose.

Procedure

Separately powder a slight excess of each crystalline substance. Weigh out the required amount of each ingredient. Mix menthol, camphor and ammonium chloride together to form a liquid. To this incorporate light magnesium carbonate, mix thoroughly and pass through a fine sieve to get a uniform powder. Pack in air tight containers, label and dispense.

Uses

This preparation is used as an insufflation to remove congestion of the nostrils.

Exercise No. 11

For :
Age :
Address :

℞

Purified talc, sterilised	50 gm
Starch, in powder	25 gm
Zinc oxide	25 gm

Label : Zinc, Starch and Talc dusting powder. Send 50 gm.
Type : Dusting powder.

Theory

This preparation contains talc which is a mineral ingredient and may be contaminated with spores of clostridium tetani and clostridium welchii (a source of tetanus). Therefore whenever talc and kaolin is to be used in dusting powders, must be sterilised by heating at 160°C for one hour to remove these micro-organisms. Purified talc has excellent flow and lubricant properties therefore is used in a number of dusting powders. Starch acts as an absorbent. Zinc oxide acts as an antiseptic and absorbs moisture.

Boric acid should no longer be used in dusting powders since it has been found that it may be absorbed in large amounts through the open skin leading to toxic reactions.

Procedure

Weigh the required quantities of purified talc, sterilised, starch, in powder and zinc oxide. Mix zinc oxide with starch, incorporate purified talc, sterilized. Mix thoroughly. Pass the mixed powders through a sieve no. 120 to remove gritty particles. After sifting, whole of the powder must again be lightly mixed. Pack the powder in siftertop containers to protect it from air, moisture and contamination as well as convenience of application.

Uses

It is used as an antiseptic dusting powder.

Marketed preparation by :

Alpine Industries, New Delhi - 110028.

Talc dusting powder

Each contains :

Purified talc	900 gm
Starch	100 gm

Exercise No. 12

For :
Age :
Address :

℞

Starch, in powder	75 gm
Zinc oxide	20 gm
Salicylic acid, in powder	5 gm

Label : Zinc and Salicylic acid dusting powder. Send 50 gm.
Type : Dusting powder.

Theory

Same as described under Exercise No. 11.

Procedure

Weigh the required quantities of each ingredient and mix them in the ascending order of their weights. Pass through sieve No. 120 to remove gritty particles. Again mix the sifted powder lightly. Pack in sifter-top container, label and dispense.

Uses

This powder is used as fungicidal dusting powder.

Marketed preparation by :
Alpine Industries, New Delhi - 110028.
Each contains :

Salicylic acid	1.5 gm
Boric acid	2.5 gm
Purified talc to	50.0 gm

Exercise No. 13

For :
Age :
Address :

℞

Tragacanth, finely powdered	15 gm
Gum acacia, finely powdered	20 gm
Starch	20 gm
Sucrose, finely powdered	45 gm

Label : The compound tragacanth powder, send 50 gm.
Type : Compound powder.

Procedure

Weigh accurately the required quantities of each ingredient. Mix them in a mortar in the ascending order of their weights. After thorough mixing, pack in wide mouth air tight container and label.

Uses

This powder is used as suspending agent.

6

Capsules

Capsules are the solid unit dosage form of medicament in which drug(s) is enclosed in a practically tasteless, hard or soft soluble container or shell made up of a suitable form of gelatin. A preservative and colouring agent may be included in the formation of shells.

There are two types of capsules (i) hard gelatin capsules which are generally used for filling the solid substances and (ii) soft gelatin or flexible capsules which are used for filling liquid and semi-solid preparations.

The hard gelatin capsules are made up of two cylindrical haves (i) longer and narrower half is called the base and (ii) shorter and wider half is called cap. The drug is filled into the base and cap is placed over it.

Hard gelatin capsules differ from soft gelatin capsules that they are hard in nature, used for filling the solid substances and are manufactured in two stages where the shells are prepared at one stage and filling is done in the second stage. The hard gelatin capsules can be extemporaneously filled and cost of production is less. Whereas soft gelatin capsules are elastic in nature due to the presence of a plastisizer in the shell, are used for filling the liquid and semi-solid preparations and filling is done in one stage only. Soft gelatin capsules cannot be prepared in the dispensary because it requires heavy machinery, hence the cost of production is more.

Enteric coated capsules

Enteric coated capsules disintegrate only in the intestine i.e. in the alkaline medium are not in the stomach where the medium is acidic in nature.

For enteric coating the filled capsules are treated with cellulose acetate phathalate or a mixture of stearic acid and butyl acetate. Now a days enteric coated capsules have been largely superseded by enteric coated tablets.

Although extemporaneous filling of capsules is rare in the dispensaries but in large hospitals it is necessary to fill such capsules for clinical trials and in cases where a particular combination of drugs is required. Hence for practice, only a few exercises are given where a small number of capsules are to be filled.

Exercise No. 14

For :
Age :
Address :

℞

Acetyl salicylic acid	300 mg
Paracetamol	150 mg
Caffeine	50 mg

Fiat capsula, Mitte tales quarta.

Sig : Unus dolore urgente sumenda.
Type : Analgesic capsules.

Procedure

Powder a slight excess of the ingredients, if already not in powder form. Weigh the calculated quantities of each ingredient. Mix them in the ascending order of their weights.

Place the mixed powders on glazed tile or glazed paper in a layer about one-third of the height of the capsule in thickness. Remove the cap from the shell. Repeatedly press the inverted base of the capsule on the powder until the capsule is filled with the desired weight of the powder. Replace the cap over the base. Check the weight by keeping an empty shell in the other pan of the balance as a tare. Fill the required number of capsules. After filling, a small amount of the powder sticks to the sides of the capsule, if not removed produce bitterness and deliquescence, thereby whole aim of filling the drug into capsules is lost. Moreover the whole process involves much handling of the capsules which leaves finger prints on the capsules. Therefore it is necessary to clean the capsules after filling. The most suitable method is to roll the filled capsules in dry towel, very lightly sprinkled with liquid paraffin. By this method sticking materials will be removed and it will also impart shine to the capsules.

Transfer the capsules in capsule vials, label and dispense.

Uses

These capsules are used as an analgesic, antipyretic and antirheumatic.

Aspirin : Analgesic, antipyretic, antirheumatic.
Paracetamol : Analgesic, antipyretic.
Caffeine : C.N.S. stimulant and it enhances the analgesic action of acetyl salicylic acid.

Marketed preparation by :

Beckem Drug International Pvt. Ltd., New Delhi - 110024.

Beckpyrin capsules
Each contains :

Aspirin	300 mg
Paracetamol	250 mg
Caffeine	30 mg

Exercise No. 15

For :
Age :
Address :

℞

Ferrous gluconate 250 mg

Fiat caps. Mitte tales sexies.

Signa : Unus post cibos sumenda.
Type : Haematinic capsules.

Procedure

Weigh the calculated amount of ferrous gluconate and fill this powder into capsules as described under exercise No. 15. Pack, label and dispense.

Uses

Ferrous gluconate is used in the prevention and treatment of iron deficiency anaemias.

Marketed preparations by :
Alpine Industries, New Delhi - 110028.
Ferrous gluconate capsules
Each contains :
Ferrous gluconate 86 mg, 325 mg, 435 mg.

Exercise No. 16

For :
Age :

℞

Chloramphenicol 250 mg

Fiat capsula. Mitte tales sexies.

Sig : Unus t.d.s.
Type : Antibiotic capsules.

Procedure

Weigh the calculated amount of chloramphenicol. Fill this powder in the capsules as described under Exercise No. 14. Pack, label and dispense.

Uses

Chloramphenicol is used as an antibiotic.

Marketed preparations by :

1. Dey's Medical Stores, Kolkata - 700087
   Chloramphenicol capsules
   Each capsule contains :
   Chloramphenicol 250 mg
2. East India Pharmaceutical Works Ltd., Kolkata - 700087
   Eastophenicol capsules
   Each capsule contains :
   Chloramphenicol 250 mg

7

Mixtures

A mixture is a liquid preparation intended for oral administration in which drug or drugs are dissolved, suspended or dispersed in a suitable vehicle and generally several doses are contained in a bottle. When only one dose is dispensed it is known as draught.

Mixtures differ from solutions that the mixtures may be homogeneous or heterogeneous and are for oral administration whereas solutions are homogeneous and are for external or internal use. Mixtures are extemporaneously prepared and they are supplied in such doses that whole of the mixture is used up within a few days. If need arises then fresh mixture is prepared.

Advantages of mixtures

1. They are more quickly effective than solid dosage forms which require previous disintegration in the body before absorption can take place.
2. Certain substances can only be given in liquid form because they are inconvenient to administer in any other form due to their liquid nature and large dose e.g. castor oil, liquid paraffin, aromatic waters etc.
3. Certain substances like potassium iodide and potassium bromide may cause pain in the stomach if given in the solid form as a powder or a tablet.
4. Certain substances are useful when they are administered in a suspension form e.g. light kaolin and bismuth salts, because in suspension form they afford large surface area for the absorption of toxic substances in the gut.
5. Mixtures are easy to administer and economical as compared to other oral preparations.

Disadvantages

1. They are comparatively less stable than solid dosage forms.
2. Incompatibility is more in liquid preparations as compared to solid ones.
3. They are more bulky and difficult to carry.

Classification

Mixtures may be classified as follows :

1. Simple mixtures.
2. Mixtures containing diffusible solids.
3. Mixtures containing indiffusible solids.
4. Mixtures containing precipitate forming liquids.
5. Mixtures containing slightly soluble liquids.
6. Miscellaneous mixtures.

1. Simple mixtures

A simple mixture is one which contains only soluble ingredients e.g. carminative mixture, diaphoratic mixture, cough expectorant etc.

Method of dispensing

(a) Dissolve the solid substances in 3/4th of the vehicle. The reasons for this is that (i) the volume occupied by the other ingredients rarely exceeds the remaining 1/4th, but if this volume exceeds, the 3/4th quantity of vehicle used must be reduced. (ii) Solution formation is hastened by using as much of the solvent as convenient.

(b) Examine the solution critically by holding the container against light. If foreign particles are viable pass the solution through cotton wool, further pour little more vehicle over cotton wool so that the solution therein is removed.

(c) Add any liquid ingredients. Volatile liquids are added at the end just before adjusting the final volume with vehicle.

(d) Add more of vehicle to produce the final volume.

(e) Transfer the mixture to a bottle, cork and thoroughly polish the bottle to remove finger prints. Attach the label, wrap the bottle and dispense.

2. Mixtures containing diffusible solids

Diffusible solids are those substance which do not dissolve in water, but on shaking they can be mixed with it and remain evenly distributed throughout the liquid for sufficient long time allowing uniform distribution of the drug in each dose. However, on standing, the insoluble solids settle at the bottom of the bottle which require re-shaking of the bottle each time whenever a dose is to be measured. Hence the bottle containing the diffusible mixture must be labelled "Shake the bottle before use". Diffusible solid include : aromatic chalk powder, bismuth carbonate, light kaolin, magnesium oxide, magnesium oxide, magnesium carbonate, magnesium trisilicate, phenolphthalein and rhubarb powder.

Method of dispensing

Finely powder the diffusible and other substances (if they are already not in fine powder) in a mortar. Mix them thoroughly. Add a small amount of vehicle out of 3/4th measured out vehicle and triturate to make a smooth cream (due to the presence of air in the interstices of many powders, they float at the surface of water and do not mix with the vehicle. To prevent this tendency, a smooth cream is prepared by adding a small amount of vehicle at first and then diluted); add the remainder of vehicle. If foreign particles are visible pass the mixture through a piece of muslin but if one or two foreign particles are visible, remove them with a glass rod. Add liquid ingredients and make up the required volume by adding more of vehicle. Transfer the mixture to a bottle, cork, polish, label and dispense. "Shake the bottle before use" label must be attached.

3. Mixtures containing indiffusible solids

Indiffusible solids are those substances which do not dissolve in water and they do not remain evenly distributed throughout the vehicle, even after shaking they immediately settle at the bottom therefore it becomes difficult to measure the dose. The indiffusible substances are made diffusible by increasing the viscosity of the vehicle by adding suitable suspending agent i.e. compound tragacanth powder or tragacanth mucilage. The former is used at the rate of 2 gm/100 ml and later at the rate of 1/4th of the volume of mixture to be prepared.

Compound tragacanth powder must be used as suspending agent when the vehicle is medicinally active because tragacanth mucilage is prepared by using chloroform water as vehicle which may replace certain amount of medicinally active vehicle thereby decreasing its medicinal and flavouring properties.

Method of dispensing

Finely powder the indiffusible solid in a mortar (if it is already not in a fine powder). Add any diffusible or soluble substances and the calculated amount of suspending agent. Mix thoroughly. Triturate the powder with a small amount of vehicle (out of 3/4th measured vehicle) to form a smooth cream then add the remainder of the measured out vehicle.

If foreign particles are present strain through muslin, add liquid ingredients and make up the required volume by adding more of the vehicle. Transfer the mixture to a bottle, label and dispense. "Shake the bottle before use" label must be attached.

Exercise No. 17

For :
Age :
Address :

℞

Ferric ammonium citrate	200 gm
Chloroform water q.s.	1000 ml

Fiat mistura. Mitte 100 ml

Sig : Cochleare modicum post cibos ter in die sumenda.
Type : Simple mixture.

Procedure

Weigh the required amount of ferric ammonium citrate and dissolve it in 3/4th of the vehicle. Filter if any foreign particles are present. Add more of vehicle to produce the required volume. Transfer the mixture to a bottle, cork, polish the bottle, label and dispense.

Uses

This mixture is used in iron deficiency anaemia.

Exercise No. 18

For :
Age :
Address :

℞

Sodium salicylate	1.0 gm
Sodium bicarbonate	1.0 gm
Sodium metabisulphite	0.01 gm
Chloroform water	to 15.0 ml

Fiat mistura. Mitte doses sexies.

Sig : Unus quartis horis sumenda.
Type : Simple mixture.

Theory

When sodium salicylate comes in contact with hydrochloric acid present in stomach, salicylic acid is formed which gets precipitated and leads to irritation in the stomach. Hence when sodium salicylate is prescribed it is usually prescribed with sodium bicarbonate which will temporarily neutralise the gastric secretion and thus minimise the formation of salicylic acid.

Sodium salicylate in solution form, specially when it is alkaline in nature, absorbs oxygen and solution becomes brownish black. Though the therapeutic value is not changed but it may lead to confusion in the mind of the patient that the mixture has spoiled and he may not like to use the mixture. Therefore to prevent the air oxidation, sodium metabisulphite is used as an antioxidant which will considerably retard the change in colour.

Procedure

Dissolve the weighed quantities of sodium salicylate and sodium bicarbonate in 3/4th of the measured out vehicle. To this dissolve sodium metabisulphite. Filter if any foreign particles are present. Incorporate more of vehicle to produce the required volume. Transfer the mixture to a bottle, cork, label and dispense.

Uses

This preparation is used as an analgesic, antipyretic and anti-rheumatic mixture.

Note : Instead of adding sodium metabisulphite, liquid extract of liquorice may be added to darken the colour of the mixture before hand. Even if any colour change due to air oxidation will be there, it will not come to the notice of the patient. Secondly the mixture may be dispensed as such with a warning to the patient about this colour change.

Exercise No. 19

For :
Age :
Address :

℞

Ammonium chloride	0.3 gm
Ammonium bicarbonate	0.2 gm
Liquid extract of liquorice	0.5 ml
Chloroform water q.s.	15.0 ml

Fiat mistura. Mitte doses sexies.

Signa : Cochleare maximum t.i.d.
Type : Simple mixture.

Procedure

Dissolve ammonium chloride and ammonium bicarbonate in 3/4th of the vehicle. Remove, if foreign particles are present. Incorporate liquid extract of liquorice. Add more of vehicle to produce the required volume. Transfer the mixture to a bottle, cork, label and dispense.

Uses

It is used as cough expectorant.

Expectorants

These are the preparations which helps in the expulsion of the sticky matter from the bronchioles, thereby relieves cough.

Cough suppressants helps in the suppression of dry cough by soothing action when there is irritation in the throat.

Ammonium chloride and ammonium bicarbonate helps in loosening the sticky matter from the bronchioles. Liquid extract of liquorice acts as a demulcent and mild expectorant. It is used to mask the nauseous taste of ammonium chloride and as a flavouring agent.

Exercise No. 20

For :
Age :
Address :

℞

Potassium bromide	8 gm
Tincture nux vomica	8 ml
Chloroform water q.s.	180 ml

Fiat mistura. Mitte 90 ml.

Signatura : Cochleare magnum ter-in die post cibos sumenda.
Type : Simple mixture.

Procedure

Dissolve potassium bromide in 3/4th of the vehicle. See against light, if foreign particles are visible, filter the solution. Add tincture nux vomica. Incorporate more of vehicle to produce the required volume. Transfer the mixture to a bottle, cork, attach the label and dispense.

Uses

It is used as sedative and anticonvulsant mixture.

Exercise No. 21

For :
Age :
Address :

℞

Magnesium sulphate	30 gm
Light magnesium carbonate	4 gm
Peppermint water to	180 ml

Fiat mistura. Mitte 90 ml.

Signa : Cochleare amplum somni hora sumenda.
Type : Mixture containing diffusible solid.

Theory

Magnesium sulphate is a soluble substance whereas light magnesium carbonate is a diffusible substance so method for mixtures containing diffusible solids will have to be followed.

Procedure

Mix the calculated quantities of magnesium sulphate and light magnesium carbonate in a mortar. Measure 3/4th of the vehicle. Out of this add small amount of vehicle to the powders. Triturate thoroughly so as to form a smooth cream. Then add remainder of the vehicle. Strain through a piece of muslim if foreign particles are present. Incorporate more of vehicle to produce the final volume. Transfer the mixture to a bottle, cork, attach the label along with "Shake the bottle before use" label and dispense.

Use

This preparation is used as a saline purgative.

Magnesium sulphate acts as purgative in large doses but in small doses it acts as an antacid. Light magnesium carbonate acts as an adjuvant and potentiates the action of magnesum sulphate. It is also used as an antacid.

Peppermint water acts as a flavouring agent.

Exercise No. 22

For :
Age :
Address :

℞

Light kaolin	2.0 gm	Light magnesium carbonate	0.5 gm
Sodium bicarbonate	0.5 gm	Peppermint water q.s.	15.0 ml

Make a mixture. Send six doses.

Direction : One dose to be taken three times a day.
Type : Mixture containing diffusible solids.

Theory

Light kaolin and light magnesium carbonate are diffusible solids and sodium bicarbonate is a soluble solid. Therefore method for mixtures containing diffusible solids will have to be followed.

Procedure

Mix the weighed quantities of light kaolin, light magnesium carbonate and sodium bicarbonate in a mortar. Measure out 3/4th of the vehicle. Out of this add a small amount of vehicle to the powder and triturate thoroughly until a smooth paste is formed. Then dilute it with remaining amount of vehicle, strain through a piece of muslin if foreign particles are present but if only one or two particles are present remove them with glass rod and avoid straining. Incorporate more of vehicle to produce the final volume. Transfer the mixture to a bottle, cork, label and attach "Shake the bottle before use" label and dispense.

Uses

It is used as an antacid mixture.

Light kaolin is purified native hydrated aluminium silicate free from gritty particles. Because it is insoluble in water, it forms a protective coating on mucous membrane of intestines, thereby decreases discomfort. It is a good intestinal absorbent. It also acts as an antacid. Therefore it is used in symptomatic treatment of enteritis, colitis and dysentry associated with food and alkaloidal poisoning.

Light magnesium carbonate is dispersed as a fine powder thereby surface area is increased. It acts as an antacid.

Sodium bicarbonate acts as an antacid as it expells out gases from the stomach. Hence it potentiates the action of light kaolin.

Marketed preparation by :

Arora Pharmaceuticals Pvt. Ltd., New Delhi - 110035.

Kaolin mixture

Each 15 ml contains :

Light kaolin	2 gm	Light magnesium carbonate	0.6 gm
Sodium bicarbonate	0.6 gm	Peppermint water	15.0 ml

Exercise No. 23

For :
Age :
Address :

℞

Calcium carbonate	0.1 gm
Light magnesium carbonate	0.1 gm
Sodium bicarbonate	0.1 gm
Simple syrup	1.0 ml
Aromatic cardamom tincture	0.1 ml
Chloroform water to	10.0 ml

Fiat mistura. Mitte 60 ml.

Sig : 10 ml to be taken three times a day.
Type : Mixture containing diffusible solids.

Note : Theory and procedure is same as described under exercise No. 22.

Uses

It is used as antacid mixture.

Exercise No. 24

For :
Age :
Address :

℞

Magnesium trisilicate	0.5 gm
Light magnesium carbonate	0.5 gm
Sodium bicarbonate	0.5 gm
Peppermint water to	10.0 ml

Fiat mistura. Mitte 60 ml.

Signa : 10 ml more dicto sumenda.

Type : Mixture containing diffusible solids.

Note : Theory and method of preparation is same as described under exercise no. 22.

Uses

It is used as an antacid mixture.

Magnesium trisilicate acts as an adsorbent and antacid. It is non-toxic even in large doses. Its action is slow but continues for a long time. Therefore it is quite useful in dyspepsia. It does not give rise to alkalosis.

Marketed preparation by :

Warner-Hindustan Division, Mumbai - 400072.

Gelusil liquid

Each 5 ml contains :

Magnesium trisilicate	625 mg
Dried aluminium hydroxide gel	312 mg

Exercise No. 25

For :
Age :
Address :

℞

Bismuth carbonate	1 gm	Prepared chalk	1 gm
Kaolin	4 gm	Tincture catechu	2 ml
Water upto	30 ml		

Fiat mistura. Mitte doses tres.

Signa : One dose to be taken three times a day.
Type : Mixture containing indiffusible solids.

Theory

Prepared chalk powder is an indiffusible solid, so method for mixtures containing indiffusible solids will have to be followed i.e. suspending agent either compound tragacanth powder in the ratio of 2 gm/100 ml or tragacanth mucilage in the ratio of 1/4th of total volume of mixture will have to be added.

Procedure

Powder the substances if they are already not in powder form. Mix the weighed quantities of bismuth carbonate, prepared chalk and kaolin in a mortar. To this incorporate the calculated quantity of compound tragacanth powder and mix thoroughly. Add a small amount of vehicle out of 3/4th measured out vehicle. Triturate so as to form a smooth cream, add remainder of the measured out vehicle. If foreign particles are present strain through a piece of muslin but if only 2-3 particles are present, remove them with glass rod and avoid straining. Add tincture catechu in the centre of the cream with continuous trituration. Incorporate more of vehicle to produce the final volume. Transfer the mixture to a bottle cork, polish the bottle, label and dispense. "Shake the bottle before use" label must be attached.

Uses

It is used as an antidiarrhoeal preparation :

1. Bismuth carbonate forms an adhesive coating over mucosa thus acts as an adsorbent. It also acts as an antacid.
2. Wherever kaolin is prescribed light kaolin must be used unless heavy kaolin is specifically mentioned. Light kaolin is purified native hydrated aluminium silicate free from gritty particles. Because it is insoluble in water, it form a protective coating on mucous membrane of intestines thereby decreases discomfort. It is a good intestinal adsorbent therefore it is used to check diarrhoea and dysentery caused by food and alkaloidal poisoning.
3. Prepared chalk is purified native calcium carbonate. It acts as an adsorbent and produces its action by covering the intestinal mucous membrane. It also acts as an antacid.
4. Tincture catechu contains catechu tannic acid which acts as an astringent. It is liberated slowly from tincture catechu hence the action remains for a long time. If pure tannic acid is used its effect is immediate but short lived, hence tincture catechu is preferred over tannic acid. It is used along with other astringents in the symptomatic treatment of diarrhoea.

Exercise No. 26

For :
Age :
Address :

℞

Acetyl salicylic acid	0.6 gm
Potassium citrate	2.0 gm
Simple syrup	4.0 ml
Chloroform water ad	30.0 ml

Fiat mistura. Mitte 90 ml.

Sig : One dose to be taken as often as necessary.
Type : Mixture containing indiffusible substance which dissolves in the presence of alkali citrates.

Procedure

Dissolve the weighed amount of potassium citrate in small amount of vehicle from measured out 3/4th of vehicle. To this incorporate acetyl salicylic acid and stir so as to dissolve it. Then add rest of the vehicle, filter if foreign particles are present. Add simple syrup and incorporate more of the vehicle to produce the required volume. Transfer to a bottle, cork, polish the bottle to remove finger prints, attach the label and dispense.

Uses

This is used as an analgesic, antipyretic mixture.

1. Acetyl salicylic acid is used as an analgesic, antipyretic and anti-rheumatic.
2. Potassium citrate is used to dissolve acetyl salicylic acid because it is insoluble in water, it dissolves due to the formation of a soluble complex. Potassium citrate is mainly used to make the urin alkaline in the treatment of inflammatory conditions of the urinary bladder and to prevent crystalluria.
3. Simple syrup is included to make the preparation sweet in taste.

Exercise No. 27

For :
Age :
Address :

℞

Sulphadimidine, in fine powder	1.0 gm
Compound tragacanth powder	0.4 gm
Orange syrup	2.0 ml
Benzoic acid solution	0.2 ml
Amaranth solution	0.1 ml
Chloroform water to	10.0 ml

Fait mistura. Mitte doses sexies.

Signa : One dose to be taken three times a day.

Type : Mixture containing indiffusible solid.

Theory

Sulphadimidine is an indiffusible solid, to make it diffusible a suspending agent will have to be included for which compound tragacanth powder is used.

Procedure

Powder the sulphadimidine if already not in fine powder. Mix it with compound tragacanth powder in a mortar. To this add orange syrup and triturate so as to form a smooth cream. As the mixture is too viscous to strain therefore remove the foreign particles with the tip of a glass rod. To this incorporate benzoic acid solution and amaranth solution previously diluted with chloroform water. Stir thoroughly so as to form a uniform mixture. Add more of vehicle to produce the required volume. Transfer the mixture to a bottle, cork, polish, attach the label and dispense "Shake the bottle before use" label must be attached.

Uses

This mixture is used in the treatment of urinary infections.

1. Sulphadimidine acts as a bacteriostatic agent. It is often the sulphonamide of choice in pneumococcal or streptococcal infections and E. coli infections of the urinary tract.
2. Compound tragacanth powder acts as a suspending agent.
3. Orange syrup is used as sweetening and flavouring agent.
4. Benzoic acid solution is used as an antibacterial and antifungal agent.
5. Amaranth solution is used as a colouring agent.

8

Emulsions

Emulsions are the biphasic liquid dosage form of medicament in which two immiscible liquids (generally one of which is water and the other is some lipid or oil) are made miscible by the addition of a third substance known as emulsifying agent or emulgent. Milk is an example of natural emulsion.

An emulsion may also be defined as a mixture of two immiscible liquids in which one liquid is dispersed as minute globules into the other. The liquid that is broken up into globules is called dispersed phase or internal phase and the liquid in which the globules are dispersed is known as continuous phase, external phase or dispersion medium. The globules remain dispersed only for a short time, separation takes place quickly upon standing. Therefore a third substance known as emulsifying agent is added to the system which forms a film around the globules of the dispersed phase thereby the globules remain scattered indefinitely in the continuous phase and a uniform, stable product is formed.

Purpose of emulsification

1. Certain medicinal agents having an unpleasant taste and odour can be made more palatable for oral administration in the form of emulsions which are otherwise difficult to take e.g. cod liver oil, shark liver oil, castor oil etc.
2. To prepare a homogeneous product containing immiscible liquids (oil and water).
3. The activity of certain drugs can be increased and action prolonged by emulsifying the drug in a suitable vehicle.
4. Sterile, stable intravenous emulsions containing fats, carbohydrates and vitamins all in one preparation can be administered to the patients who are unable to take these vital substances by oral route.
5. Dermatological preparations like liniments, lotions and creams help in quick absorption of drugs from skin surface when applied externally.

Types of emulsions

There are two types of emulsions :

1. Oil in water type (o/w) emulsions.
2. Water in oil type (w/o) emulsions.

In oil in water type emulsions the oil is in the dispersed phase whereas water is in the continuous phase. These types of emulsions are prepared by using emulsifying agents generally obtained from natural sources like gum acacia, tragacanth, methyl cellulose, saponins and soaps formed from monovalent bases like Na^+, K^+ and NH_4^+, oil in water type emulsions are preferred for internal use because the unpleasant taste and odour is masked by emulsification and oil being in a finely dispersed state is more quickly assimilated in the body. Some of the o/w type emulsions can also be used externally.

In water in oil type emulsions, the water is in the dispersed phase whereas oil is in the continuous phase. These types of emulsions are mainly used externally. Emulsifying agents like wool fat, resins, bees wax, synthetic compounds and soaps formed from divalent bases like Ca^{++}, Mg^{++} and Zn^{++} are used for preparation of water in oil emulsions.

Tests to identify the type of emulsions

Since both types (o/w) and (w/o) of emulsions are similar in appearance therefore it is very difficult to differentiate them with naked eye. They can be identified with the help of following tests.

(a) Dilution test

Take a few drops of emulsion in a test tube and dilute it with 2-3 drops of water and shake. If the emulsion remains uniform, it is o/w emulsion but if the water forms a separate layer it is w/o emulsion. Reverse will be the case if the emulsion is diluted with oil.

(b) Dyesolubility test

Mix an oil soluble dye like scarlet red with an emulsion. Place a drop of it on microscope slide and see under the microscope, if the continuous phase appears to be red, it is w/o emulsion but if scattered globules appears red and continuous phase colourless, it is o/w emulsion. The test can be repeated by using amaranth, a water soluble dye. If the continuous phase appears red it is o/w emulsion but if scattered globules appears red and continuous phase colourless it is w/o emulsion.

(c) Conductivity test

Take emulsion in a beaker and dip a pair of electrodes in it which are connected through a low voltage lamp. Pass the current through the electrodes. If the bulb glows, it is o/w emulsion but if it does not glow then the emulsion is w/o because in the first case water is in the continuous phase which has allowed the current to pass through but in the second case oil is in the continuous phase through which the current has not passed because water is good conductor of electricity whereas oil is a bad conductor of electricity.

Emulsifyig agents

Emulsifying agents or emulgents are the substances which reduce the interfacial tension between the two phases i.e. aqueous phase and oily phase thus make them miscible with each other and form a stable emulsion.

No single emulsifying agent possess all the properties required for the preparation of stable emulsion therefore sometimes it becomes necessary to use two or more than two emulsifying agents instead of one to get a product of desired qualities. Emulsifying agents commonly used include :

1. Natural emulsifying agents from vegetable sources : Acacia, tragacanth, agar, chondrus, pectin and starch.
2. Natural emulsifying agents from animal sources : Gelatin, egg yolk and wool fat.
3. Semi-synthetic polysaccharides; Methyl cellulose and sodium carboxymethyl cellulose.
4. Synthetic emulsifying agents : Anionic, cationic and non-ionic.
5. Inorganic emulsifying agents : Milk of magnesia, magnesium oxide, magnesium trisilicate, magnesium aluminium trisilicate, bentonite etc.

Preparation of emulsions

Emulsions can be prepared by the following methods :

1. Dry gum method.
2. Wet gum method.
3. Bottle method.

In dry gum method the oil is triturated with gum and then water is added to make a primary emulsion whereas in the wet gum method the gum is triturated with water to form a mucilage first and then oil is added for the preparation of primary emulsion.

For extemporaneous compounding of emulsions by dry gum method and wet gum method the most efficient apparatus used is mortar and pestle. Wedge wood or porcelain mortar and pestle is used which should be flat bottomed and rough on the inner surface so as to produce fine particles of the dispersed globules. Glass mortars should not be used because of their smooth surface which will not produce a good emulsion.

The quantities of oil, gum and water required for primary emulsion are as follows :

Proportion of	Oil	:	Water	:	Gum
Fixed oils	4	:	2	:	1
Volatile oils	4	:	4	:	2

The most commonly used fixed oils include castor oil, code liver oil, shark liver oil, olive oil, almond oil and liquid paraffin (mineral oil).

The volatile oils include turpentine oil, sandal wood oil, cinnamon oil and peppermint oil.

1. Dry gum method

The method is also known as 4 : 2 : 1 method because these figures represent the proportion of oil, water and gum acacia required for the preparation of primary emulsion.

Measure the given quantity of oil with a clean and dry measure and transfer it to a dry mortar. To this add the calculated quantity of acacia and triturate rapidly so as to mix them. To this incorporate water required for primary emulsion and triturate rapidly without ceasing till a clicking sound is produced and the product becomes white or nearly white. At this stage the emulsion is known as primary emulsion. Then add more of water to produce the required volume. If any soluble ingredient is also to be incorporated that must be dissolved in the second part of water to be added after making the primary emulsion and to produce the final volume.

2. Wet gum method

The proportion of oil, water and gum are same as for dry gum method. In this method the calculated quantity of gum is triturated with water required for primary emulsion, to form a mucilage. Then the given amount of oil is incorporated in small portions with rapid titration until a clicking sound is produced and the product becomes white or nearly white. Add more of vehicle to produce the final volume.

3. Bottle method

The bottle method is used for the preparation of emulsions of volatile and other non-viscous oils. The emulsions can be prepared by both dry gum and wet gum methods. Because of low viscosity the volatile oils require greater amount of gum for emulsification than fixed oils.

In the preparation of majority of emulsions it is necessary to prepare first the primary emulsion which is diluted afterwards with more of vehicle. It is very difficult to mix whole of the oil with whole of water all at once because the volume of the liquid becomes quite large and the shearing force required to cut the dispersed phase into suitable size of globules is difficult to obtain. However, if the oil is first emulsified with only a small portion of water and emulsifying agent, a primary emulsion is formed by suitably reducing the size of oil globules, which can be diluted afterwards with more of water. The entire process of emulsion formation depends on the proper preparations of primary emulsions. Better the primary emulsion formed better will be the emulsion.

Creaming of emulsion

In creaming of emulsion the dispersed globules of oily phase collect at the surface of emulsion and form a thick layer over there whereas in sedimentation the dispersed globules move downward towards the bottom and form a layer over there. Creaming is a temporary phase because it can be re-distributed by mild shaking or stirring to get a homogeneous product. At the same time creaming is undesirable because a badly creamed emulsion may lead to cracking with complete separation of two phases. The rate of creaming is governed by Stokes law which states :

$$V \propto \frac{2r^2 (d_1 - d_2) g}{9\eta}$$

The factors which lead to creaming include radius of globules, difference between the densities of dispersed phase and continuous phase, gravitational constant and viscosity of the dispersion medium.

Cracking of emulsion

In cracking of emulsions the coalescence of the dispersed globules take place and two separate layers of the dispersed phase and continuous phase are formed which are difficult to redisperse by shaking or stirring to get the original product. Hence cracking is more serious than creaming. The factors which lead to cracking include : addition of emulsifying agent of opposite type, decomposition or precipitation of emulsifying agent, addition of common solvent, micro-organisms, high temperature and creaming.

Exercise No. 28

For :
Age :
Address :

℞

Castor oil	8 ml
Water	ad 30 ml

Fiat emulsio.

Signa : More dicto sumenda.
Type : O/w type emulsion for internal use.

Theory

Castor oil is a fixed oil and is not miscible with water. To make it miscible a third substance known as emulsifying agent in the ratio of 4 : 2 : 1 i.e. oil : water : gum will be used for the preparation of primary emulsion. Gum acacia will be used as emulsifying agent because emulsions prepared with gum acacia remain stable for sufficient long time.

Formula for primary emulsion :

Oil	:	Water	:	Gum
4	:	2	:	1
8 ml	:	4 ml	:	2 gm

Procedure

Wet gum method

Thoroughly clean and dry a pestle and mortar. Weigh out 2 gm gum acacia and transfer it to the mortar. Measure 4 ml water and triturate it with gum so as to form a mucilage. To this add 8 ml castor oil in small quantities at a time with thorough trituration after each addition. Triturate briskly without ceasing until a clicking sound is produced and the product becomes white or nearly white. At this stage the emulsion is known as primary emulsion. Add about 10 ml more of vehicle in small quantities at a time with constant trituration so as to get a homogeneous product.

Transfer the emulsion to a measure, add more of vehicle to produce the final volume 30 ml, stir thoroughly so as to form a uniform emulsion. Transfer the preparation to a bottle, cork, polish the bottle to remove finger prints, label and dispense. Attach the secondary label "Shake well before use".

Uses

Castor oil is used as purgative.

Precautions

Because of its prompt action castor oil should not be administered at bed time, preferably it should be given early in the morning.

Exercise No. 29

For :
Age :
Address :

℞

Olive oil	30 ml
Water	to 120 ml

Fiat mistura. Mitte 60 ml

Signa : Cochleare medium bis in die sumendum.
Type : O/w type emulsion for internal use (olive oil is a fixed oil).

Note : Theory and method of preparation is same as described in castor oil emulsion.

Uses

Olive oil emulsion may be given as a nitrogen-free diet during the treatment of renal failure. Olive oil also acts as a nutrient, demulcent and mild laxative.

Exercise No. 30

For :
Age :
Address :

℞

Liquid paraffin	30 ml
Water	q.s. 100 ml

Fiat emulsion. Mitte 50 ml.

Sig : Cochleare amplum hora somni sumenda.

Type : O/w type emulsion for oral use (liquid paraffin is a mineral oil which will be treated as fixed oil for the preparation of emulsions).

Note : Theory and method of preparation is same as described under castor oil emulsion.

Uses

Liquid paraffin emulsion is used as laxative and acts as emollient purgative in chronic constipation especially during pregnancy and old age.

Marketed preparation by :

Rays Laboratories Pvt. Ltd., Kolkata - 700007.

Phenolax liquid

Each 5 ml contains :

Liquid paraffin	1.25 ml
Phenolphthalein	17.0 mg
Magnesium hydroxide	100 mg

Exercise No. 31

For :
Age :
Address :

℞

Liquid paraffin	60.0 ml
Acacia	15.0 gm
Phenolphthalein	2.0 gm
Syrup	15.0 ml
Agar	1.5 gm
Cinnamon water to	180.0 ml

Make an emulsion. Send 90 ml.

Sig : One table spoonful to be taken twice a day.
Type : O/w type emulsion containing oil and water insoluble substance i.e. phenolphthalein.

Theory

Phenolphthalein is insoluble in liquid paraffin and water therefore it will have to be finely powdered if already not in fine powder, then mix with acacia before adding liquid paraffin and water to get primary emulsion. The final emulsion will have to be passed through a homogeniser to further reduce the particle size to get a stable and whiter emulsion because the ordinarily prepared liquid paraffin emulsions are very coarse in nature.

Formula for primary emulsion :

Liquid paraffin	60 ml
Acacia	15 gm
Phenolphthalein	2 gm
Cinnamon water	30 ml

Procedure

Finely powder phenolphthalein in a mortar. Mix it with acacia, add liquid paraffin and triturate. Incorporate cinnamon water required for primary emulsion, little at a time with constant trituration until a clicking sound is produced and a white product is formed. This will constitute the primary emulsion, the volume of which will be approximately 107 ml. Separately in a tared dish dissolve agar in about 60 ml water by gentle heat; while hot adjust the volume to 60 ml because during heating small quantity of water will evaporate, therefore the volume is adjusted. Add syrup and cinnamon water, stir so as to get a uniform mass. Gradually add the hot agar solution to the warm primary emulsion with constant trituration until uniform. Pass the product through a homogeniser in order to further reduce the particle size of the oil globules to get a stable and whiter product. Pass sufficient vehicle

through the homogeniser to produce the required volume. Transfer to a bottle, label and dispense. Attach the secondary label, "Shake the bottle before use".

Uses

This preparation is used as purgative. (1) Phenolphthalein is an irritant purgative which is usually given at night to act in the morning. It is administered alone or along with other purgatives such as liquid paraffin which enhances its action. (2) Acacia acts as an emulsifying agent but because of low viscosity a stable emulsion is not produced. To increase the viscosity agar is used as secondary emulsifying agent. Only a very small proportion of agar in solution can be included in an emulsion. More than 1% agar will produce a solid or semi-solid preparation, so agar must be used in proper proportion only along with primary emulsifier. (3) Syrup acts as a sweetening agent and cinnamon water acts as flavouring agent.

Marketed preparation by :

Warner-Hindustan Division, Mumbai - 400072.

Agarol emulsion. Each 30 ml contains :

Liquid paraffin	9.54 ml
Phenolphthalein	400 mg
Agar	60 mg

Emulsifying agents and excipients to 30 ml in aqueous base containing glycerin and sorbitol solution.

Exercise No. 32

For :
Age :
Address :

℞

Cod liver oil	30 ml
Syrup	12 ml
Ferric ammonium citrate	4 gm
Cinnamon water q.s.	90 ml

Fiat emulsio. Send 45 ml.

Signa : Cochleare minimum bis in die sumenda.

Type : O/w type emulsion for internal use.

Note : Theory is same as described under castor oil emulsion.

Formula for primary emulsion :

Cod liver oil	30.0 ml
Acacia powder	7.5 gm
Cinnamon water	15.0 ml

Procedure

Dry gum method

Measure the required amount of cod liver oil with a dry measure. Place it in a dry mortar and add gum acacia powder, mix thoroughly. Add measured amount of cinnamon water little at a time with continuous trituration until a white product is formed. At this stage volume of primary emulsion will be about 52.5 ml.

Dilute the syrup with about 20 ml of cinnamon water. To this dissolve ferric ammonium citrate. Add this solution little at a time to primary emulsion with continuous and brisk trituration till whole of the ferric ammonium citrate solution has been added. Transfer it to a measure, rinse the mortar with a little quantity of cinnamon water, transfer the rinsings to the contents of the measure. Add more of vehicle to produce the required volume. Transfer the preparation to a bottle, cork, polish, label and dispense. Attach "Shake the bottle before use" label.

Uses

1. Cod liver oil is a valuable source of vitamin A and D and of readily digestible fat. It is used as a dietary supplement for infants and children to prevent the occurrence of rickets and to improve nutrition and calcification of bones in undernourished children and in patients with rickets.
2. Ferric ammonium citrate is added as an antianaemic drug and help in patients who are suffering from iron deficiency anaemia.
3. Syrup acts as a sweetening agent and chloroform water is used as flavouring agent.

Exercise No. 33

For :
Age :
Address :

℞

Fish liver oil	15 ml
Egg yolk	2 ml
Simple syrup	2 ml
Chloroform water ad	50 ml

Fiat emulsio.

Sig : Cochleare minimum post cibos sumenda.
Type : O/w type emulsion.

Note : Theory and uses are same as described under cod liver oil emulsion. Egg yolk is used as an emulsifying agent (Protein type). It also acts as a nutrient.

Formula for primary emulsion :

Fish liver oil	15 ml
Egg yolk	2 ml
Chloroform water	4 ml

Procedure

Carefully separate egg yolk (yellow part of egg) from egg white. Measure 2 ml of egg yolk and transfer it to a thoroughly clean and dry mortar. Measure 15 ml of fish liver oil in a dry measure and add gradually to the mortar with trituration. Add 4 ml of chloroform water little at a time to the mortar with continuous trituration until a white product is formed. Add simple syrup and mix. Transfer the contents of the mortar to a measure, rinse the mortar, add the rinsing to the measure. Incorporate more of vehicle to produce the required volume. Transfer the preparation to a bottle, cork, polish, label and dispense. Attach "Shake the bottle before use" label.

Precautions

Emulsions prepared with egg yolk must be freshly prepared and consumed within 2-3 days after its preparation because it undergoes putrefaction on storage. Therefore emulsions so prepared should be kept in a refrigerator. Egg yolk is rarely used in industrial preparations because the emulsions so prepared get spoiled during transportation, therefore, it is mainly used in extemporaneous preparations meant for internal use.

Exercise No. 34

For :
Age :
Address :

℞

Calciferol solution	0.15 ml	Glycerin	0.3 ml
Water ad	5.0 ml		

Fiat emulsio. Mitte 50 ml.

Signa : 5 ml to be taken daily.
Type : O/w type emulsion for internal use containing a small volume of oily substance.

Theory

In this prescription the quantity of calciferol solution (an oily liquid) prescribed is very small i.e. about 3 percent. Generally speaking, emulsions containing appreciately less than 20% of oily liquid and prepared with the usual proportion of gum acacia, become unstable and readily cream. Therefore to prevent creaming, a bland of fixed oil (e.g. arachis, almond or alive) should be added to raise the total quantity of oily liquid to approximately 20%.

The substances which are prescribed in this way and require previous dilution with a fixed oil before emulsification include calciferol solution, concentrated solution of vitamin A, concentrated solution of vitamin A and D, concentrated solution of vitamin D, halibut liver oil and bromoform.

Formula for primary emulsion :

Calciferol solution	1.5 ml	Olive oil	8.5 ml
Powdered acacia	2.5 gm	Water	5.0 ml

Procedure

Dry gum method

Take the calculated quantities of calciferol solution and olive oil in a dry mortar, add acacia powder and mix thoroughly. To this add measured amount of water, little at a time with continuous trituration until a white product is obtained and a clicking sound is produced. At this stage the emulsion is known as primary emulsion.

To the primary emulsion add glycerin with trituration. Incorporate more of vehicle to dilute the emulsion. Transfer to a measure, rinse the mortar with small quantity of vehicle, add the rinsings to the measure. Incorporate more of vehicle to produce the required volume. Transfer the preparation to a bottle, cork, polish, label and dispense. Attach "Shake the bottle before use" label.

Uses

This emulsion is used in vitamin D deficiency which may lead to rickets in children and osteomalacia in adults.

Precautions

Excessive use of calciferol may produce anorexia, nausea, vomiting, diarrhoea, loss of weight, headache, vertigo and eventually a raised blood urea. Overdoses of calciferol in children is dangerous therefore liquid preparations of calciferol for infants and children should particularly be carefully measured.

9

Semisolid Dosage Forms

OINTMENTS

Ointments are the soft semisolid preparations meant for external application to the skin or mucous membrane. They usually contain a medicament or medicaments dissolved, suspended or emulsified in the base. Ointments are used for their emollient and protective action to the skin. They may also be used as vehicles or bases for the topical application of medicinal substances. An ointment for use in the eyes is known as ophthalmic ointment.

Ointment bases

An ointment base is that substance or part of an ointment which serves as a carrier or vehicle for the medicament. An ideal ointment base should be inert, stable, smooth, compatible with skin, non-irritating and should release the incorporated medicament readily. Since there is no single ointment base available which possesses all these qualities, therefore, it becomes necessary to use more than one ointment base in the preparation of ointments.

Classification of bases

1. Oleaginous bases

These bases are insoluble in water, non-washable and greasy in nature. Examples are Petrolatum (yellow soft paraffin and white soft paraffin), liquid paraffin and hard paraffin which are obtained from petrol. The animal fat includes lard.

2. Absorption bases

They are generally anhydrous bases which can absorb a large amount of water but still retain their ointment like consistency. These are also greasy in nature but can be easily removed from the skin e.g. wool fat (lanolin), wool alcohols, bees wax, cholesterol etc.

3. Emulsion bases

These are semisolid o/w and w/o emulsions. Some additional amount of water can be incorporated in both the types and still retain soft cream like consistency e.g. hydrophilic ointment etc.

4. Water soluble basis

These bases contain only the water soluble ingredients and not fats or greasy substances that is why sometimes they are known as greaseless bases e.g. carbowaxes.

Preparation of ointments

Ointments can be prepared by two methods :

1. Trituration method
2. Fusion method

1. Trituration method

Powder the medicament if already not in fine powder. Triturate it with a small amount of base on an ointment slab with a stainless steel spatula with long blade. Incorporate this to the rest of the base with thorough trituration until uniform. If liquids are also to be incorporated, pestle and mortar should be used for the purpose.

2. Fusion method

When an ointment base contains a number of solid substances, melt them in decreasing order of their melting points to avoid over-heating of low melting point substances. Add the medicament to the melted bases and stir thoroughly until the mass cools down and a homogeneous product is formed.

If any liquid or aqueous substance is also to be incorporated, that must be heated to about the same temperature as the melted bases. After mixing the two portions they should be stirred uniformly and thoroughly until a homogeneous mass is obtained. Rapid cooling should be avoided.

Creams

Creams are thought of as ointments but usually contain a water soluble base due to which they can be easily removed from the skin. They are of softer consistency and have a lighter body than true ointments. When applied to the skin, creams leave no visible evidence of their presence on the skin.

Pastes

Pastes are the semi-solid preparations meant for external application to the skin. They differ from ointments that they generally contain a large amount of finely powdered solids such as starch, zinc oxide, calcium carbonate etc. They are quite thick and stiff than ointments but are less greasy than ointments.

Packing

Ointments, creams and pastes should be packed in ointment jars or collapsible tubes. They should be labelled with good quality of paper and adhesive. The label should be attached to the collapsible tubes towards the top because during use the tube is rolled up and this will prevent the spoilage of the label until the tube is practically empty. A secondary label "for external use only" must be attached.

Exercise No. 35

For :
Age :
Address :

℞

Calamine, finely sifted	15 gm
White soft paraffin	85 gm

Label : The calamine ointment.

Procedure

Pass the calamine through a very fine sieve to get a fine powder. Triturate the calamine with a portion of white soft paraffin on an ointment slab with ointment spatula, until smooth. Gradually add the remainder of white soft paraffin with continuous trituration until a uniform ointment is obtained.

Pack the ointment in a wide mouth container or in an ointment jar, label and dispense. The container must be labelled with directions "For external use only."

Uses

Calamine has a mild astringent action on the skin and is used in ointments to relieve discomfort of dermatitis.

Marketed preparation by :
Alpine Industries, New Delhi - 110028.
Calamine ointment
Each contains :

Calamine	150 gm
White soft paraffin	850 gm

Exercise No. 36

Prepare 100 gm of simple ointment.

Formula

Wool fat	50 gm
Hard paraffin	50 gm
Cetostearyl alcohol	50 gm
White soft paraffin	850 gm

Type : Absorption ointment base.

Procedure

Melt hard paraffin and cetostearyl alcohol on water bath. To this incorporate wool fat and white soft paraffin. Stir until all the ingredients are melted.

Examine the contents for foreign particles. Decant or strain, if necessary. Stir the mixture thoroughly until cold. Pack in suitable container and label.

Uses

Simple ointment is used as an ointment base.

Note :

1. Instead of white soft paraffin, yellow soft paraffin can be used for a product with highly coloured ingredients.
2. All the above mentioned ingredients can be melted together but it is advisable to melt them in the decreasing order of their melting points because in this way the ingredients will be melted in lesser time moreover overheating of the ingredients will be avoided.

Exercise No. 37

For :
Age :
Address :

℞

Precipitated sulphur, finely sifted	10 gm
Simple ointment prepared with white soft paraffin	90 gm

Label : The ointment. Send 50 gm.
Sig : Wash the skin thoroughly with warm soap water and rub the ointment on the affected skin.

Procedure

Triturate the precipitated sulphur with small amount of simple ointment on an ointment slab. Gradually add the remainder of the simple ointment with thorough trituration until a homogeneous mass is obtained. Pack in an ointment jar, label and dispense.

Uses

Sulphur ointment is used as fungicidal ointment and is also used in scabies and ring worm infections.

Marketed preparation by :
Medo Chem. Labs. Pvt. Ltd., Delhi-110032.
Sulphur ointment
Each contains :

Sulphur sublimed	100 gm
Simple ointment	900 gm

Exercise No. 38

For :
Age :
Address :

℞

Zinc oxide, finely sifted	15 gm
Simple ointment	85 gm

Label : The ointment, send 50 gm.

Procedure

Triturate the zinc oxide with a portion of simple ointment on an ointment slab until smooth. Gradually add the remainder of the simple ointment with thorough trituration until a homogeneous mass is obtained. Pack in an ointment jar, label and dispense.

Uses

Zinc oxide is used in ointments as a mild astringent for the skin, as a soothing and protective application in eczema.

Marketed preparations by :

1. Bell Pharma Pvt. Ltd., Mumbai - 400016.
   Bell cream
   Contains :
   Zinc oxide 2%
2. Agrawal Pharmaceuticals, Delhi - 110092.
   Zinc oxide ointment
   Each contains :
   Zinc oxide 0.150 gm
   Simple ointment 0.850 gm

Exercise No. 39

For :
Age :
Address :

℞

Salicylic acid, in fine powder	3 gm
Benzoic acid, in fine powder	6 gm
Emulsifying ointment	91 gm

Label : The Whitfield's ointment, send 50 gm.
Directions : To be applied on the affected parts.

Procedure

Separately pass the salicylic acid and benzoic acid through No. 120 sieve. Mix the required quantities lightly on a tile. Triturate the mixed powders with a portion of emulsifying ointment until smooth. Gradually incorporate the remainder of the emulsifying ointment with thorough trituration until smooth. Transfer to an ointment jar, label and dispense.

Uses

It is used as fungicidal ointment.

Marketed preparations by :

1. Dermocare Labs, Ahmedabad - 380001
   Sallderm skin ointment
2. Jilichem Labs (India) Ltd., Ahmedabad - 382445
   Whitfield's ointment
3. Tuton Pharmaceuticals, Ahmedabad - 382330
   Each contains :
   Salicylic acid 3% w/w
   Benzoic acid 6% w/w
   Paraffin base sufficient quantity.

Exercise No. 40

Prepare 50 gm Wool Alcohols Ointment.

Formula

Wool alcohols	3 gm
Hard paraffin	12 gm
White soft paraffin	5 gm
Liquid paraffin	30 gm

Type : Absorption ointment base.

Procedure

Melt the required quantities of wool alcohols, hard paraffin, white soft paraffin and liquid paraffin in a dish and stir until cold and a uniform mass is obtained.

Uses

It is used as an ointment base.

Note :

1. The quantities of hard paraffin, liquid paraffin and soft paraffin may be varied to produce a base of desired consistency.
2. For highly coloured drugs wool alcohols ointment may be prepared by using yellow soft paraffin instead of white soft paraffin.

Exercise No. 41

For :
Age :
Address :

℞

Salicylic acid, finely sifted	0.2 gm
Wool alcohols ointment	9.8 gm

Fiat : Ungentum.
Sig : More dicto applicandus.

Procedure

Melt the wool alcohols ointment, gradually add the salicylic acid and stir thoroughly until cold and homogeneous mass is formed. Pack in suitable container, label and dispense.

Uses

It is used as bacteriostatic and fungicidal ointment.

Marketed preparation by :

Alpine Industries, New Delhi - 110028.

Salicylic acid ointment
Each contains :

Salicylic acid	20 gm
Wool alcohol	980 gm

Exercise No. 42

For :
Age :
Address :

℞

Cetrimide	1 gm
Cetostearyl alcohol	10 gm
White soft paraffin	10 gm
Liquid paraffin	29 gm
Purified water	50 gm

Make a cream. Send 25 gm.

Label : The antiseptic cream.
Type : Cream prepared by fusion method.

Procedure

Melt cetostearyl alcohol, white soft paraffin and liquid paraffin together. Separately dissolve cetrimide in purified water and warm it to almost same temperature (about 60°C) as that of melted substances. Add the warmed aqueous liquid to the melted mixture and stir thoroughly until cold. Pack is suitable container, label and dispense.

Uses

It is used as an antiseptic cream for the treatment of wounds and burns; for the pre-operative cleansing of the skin and for the removal of scabs and crusts in skin diseases.

Cetrimide is a quaternary ammonium compound. It is relatively non-toxic antiseptic with detergent properties. It is quite effective against Gram-positive micro-organisms but less effective against Gram-negative micro-organisms. Aqueous solutions containing 0.1 to 1.0% cetrimide are quite commonly used in hospitals.

Marketed preparations by :
Alpine Industries, New Delhi - 110028.
Cetrimide cream
Each contains :

Cetrimide	5 gm
Cetostearyl alcohol	50 gm
Purified water	445 gm
Liquid paraffin	500 gm

Exercise No. 43

For :
Age :
Address :

℞

Proflavin	0.05 gm
Emulsifying wax	5.0 gm
White soft paraffin	10.0 gm
Liquid paraffin	5.0 gm
Purified water	30.0 gm

Label : The proflavine antiseptic cream. Send 25 gm.

Procedure

Melt together emulsifying wax, white soft paraffin and liquid paraffin. Separately dissolve proflavin in purified water and warm to almost same temperature (about 60°C) as that of melted paraffins. Add this warmed solution to the melted bases and stir thoroughly until cold and uniform mass is formed. Pack in a suitable container, label and dispense.

Uses

It is used as an antiseptic cream.

Marketed preparation by :

Alpine Industries, New Delhi - 110028.

Acriflavine ointment

Each contains :

Acriflavine	0.1% w/w

Exercise No. 44

For :
Age :
Address :

℞

Liquid paraffin	20.0 gm
Hard paraffin	4.5 gm
Lanette wax	3.5 gm
Glycerin	4.5 gm
Water	17.5 gm
Propyl paraben	0.1 gm
Perfume	Sufficient quantity

Label : The cold cream.

Procedure

Melt the lanette wax, hard paraffin, and liquid paraffin on water bath. Separately dissolve propyl paraben in water, add glycerin to it and heat this solution to almost same temperature (about 60°C) as that of melted bases. Mix both the solutions and stir continuously until cold and uniform product is obtained. Pack in suitable container, label and dispense.

Uses

It is used as skin protective and skin smoothener.

Lanette wax was originally a German patent and was widely used in dermatological preparations. It consists of a mixture of sodium cetostearyl sulphate (10 per cent) and cetyl and stearyl alcohols (90 per cent). For pharmaceutical purposes it is now largely replaced by emulsifying wax which is suitably prepared by mixing 90 gm of cetostearyl alcohol, 10 gm of sodium lauryl sulphate and 4 ml of water.

Lanette wax produces an oil in water emulsion. Propyl paraben acts as a preservative. Glycerin is used for water retaining and emollient properties.

Exercise No. 45

For :
Age :
Address :

℞

Heavy Kaolin, finely sifted, dried at 100°C	52.7 gm
Boric acid, finely sifted	4.5 gm
Methyl salicylate	0.2 ml
Thymol	0.050 gm
Peppermint oil	0.05 ml
Glycerin	42.5 gm

Fiat Cataplasma. Mitte 50 gm.

Sig : Apply to the affected part.

Procedure

Mix heavy kaolin and boric acid with glycerin. Heat at 120°C for one hour with occasional stirring. Allow to cool. Separately dissolve thymol in methyl salicylate and peppermint oil, add to the cooled mixture with thorough mixing until homogeneous. Transfer to a tightly closed container, label and dispense.

Theory

Heating to 120°C destroys the Clostridium tetani present in kaolin. It is resistant to low heat. Too much heating should be avoided which may decompose glycerin to form acrolin. A suitable tightly closed container should be used to prevent absorption of moisture by glycerin and loss of volatile substances.

Uses

It is used as counter-irritant and anti-inflammatory agent.

Exercise No. 46

For :
Age :
Address :

℞

Zinc oxide, finely sifted	25 gm
Starch, finely sifted	25 gm
White soft paraffin	50 gm

Make a paste. Send 25 gm.

Label : Spread thickly on white lint and apply to the affected area.

Type : Paste with semi-solid base prepared by fusion and trituration method.

Procedure

Separately pass the zinc oxide and starch through sieve No. 120. Melt the white soft paraffin on water bath.

Mix the required weight of powders in a warm mortar. Add small amount of melted base with continuous trituration until smooth. Gradually add remainder of the base and mix until cold and uniform paste is obtained. Pack in ointment jar, label and dispense.

Uses

It is used as an antiseptic paste.

Note :

1. Warm mortar is used because it is much easier to manipulate the paste in warm mortar.
2. This preparation is also known as compound zinc paste.

Exercise No. 47

For :
Age :
Address :

℞

Starch, finely sifted	24 gm
Zinc oxide, finely sifted	24 gm
Salicylic acid, finely sifted	2 gm
White soft paraffin	50 gm

Make a paste. Send 25 gm.

Type : Paste with semisolid base prepared by fusion method.

Procedure

Melt the white soft paraffin on water bath. Incorporate the starch, zinc oxide and salicylic acid which has been previously sifted through sieve No. 120. Stir until cold and homogeneous paste is obtained. Pack, label and dispense.

Uses

It is used as an antifungal paste.

Note : This paste is also known as Lassar'a paste.

Marketed preparation by :

Alpine Industries, New Delhi - 110028.

Zinc and salicylic acid paste (Lassar's Paste)

Formula is same as given above.

Exercise No. 48

For :
Age :
Address :

℞

Resorcinol, finely sifted	1.25 gm
Precipitated sulphur	1.25 gm
Zinc oxide, finely sifted	10.00 gm
Emulsifying ointment	12.50 gm

Fiat : Pasta. Mitte 50 gm.

Type : Paste containing semi-solid base and prepared by trituration method.

Procedure

Pass separately the resorcinol, precipitated sulphur and zinc oxide through sieve No. 120 and mix the weighed quantities of these substances with a portion of emulsifying ointment until smooth. Gradually add the remainder of the emulsifying ointment with thorough trituration until smooth. Pack in suitable container, label and dispense.

Use

It is used in the treatment of psoriasis (chronic skin disease in which red scaly patches develops), eczema and other skin diseases.

Exercise No. 49

For :
Age :
Address :

℞

Glycerin, previously, heated at 120° for one hour and cooled	55.0 gm
Phenol	0.5 gm
Dried magnesium sulphate	A sufficient quantity

Label : The paste.
Sig : To be applied to the boil as directed.
Type : Paste with liquid vehicle prepared by trituration.

Procedure

Dry about 70 gm of the dried magnesium sulphate at 150° for 1½ hour or at 130° for 4 hours and allow to cool in a desiccator. Mix 45 gm of this dried material in a warm mortar with glycerin in which phenol has already been dissolved. Mix thoroughly until homogeneous paste is obtained. Pack in well closed glass container, label and dispense.

Use

This paste is used to draw out the contents of boils and carbuncles.

Note :

1. This paste is also known as Morison's paste.
2. This paste should be free from micro-organisms because it is used to draw out the contents of boils and carbuncles and must not cause re-infection.
3. Micro-organisms in the magnesium sulphate are destroyed by the drying process.
4. The phenol has mild analgesic and antiseptic action. Although in the presence of glycerin the anti-microbial activity of phenol is decreased but it helps to maintain sterility of the preparation.
5. Magnesium sulphate powder sediments quickly, therefore, this paste should not be packed in collapsible tubes. The container should be labelled with the instructions that the contents should be stirred well before use.

Marketed preparation by :
Agrawal Pharmaceuticals, New Delhi - 110092.
Magnesium sulphate ointment
Each 100 gm contains :

Dried magnesium sulphate	64 gm
Glycerin	35.5 gm
Phenol	0.5 gm

10

Monophasic Liquid Dosage Forms

Syrups

Syrups are sweet, viscous, concentrated aqueous solutions of sucrose or other sugars in water or any other suitable aqueous vehicle. When purified water alone is used in making the solution of sucrose, the preparation is known as simple syrup. When the preparation contains some medicinal substance it is known as medicated syrup. When the syrup does not contain any medicament but contains various aromatic or pleasently flavoured substances are known as flavouring syrups. They are used for masking the disagreeable taste of bitter or saline drugs. They are also used as vehicles or flavours for extemporaneous preparations.

Simple syrup is prepared by dissolving 66.7 gm sugar in sufficient amount of water to produce 100 gm. Recently prepared syrups should be used. If the syrup is to be stored for sometime it must be preserved by adding suitable preservative.

Gargles

Gargles are aqueous solutions used for the treatment of an infection of the throat. Usaually they are concentrated solutions and must be diluted with water before use. In using the gargles they are brought into intimate contact with the mucous membrane of the throat and are allowed to remain there for a few moments after which they are thrown out of the mouth.

Mouthwashes

A mouthwash is an aqueous solution with a pleasant taste and odour used for rinsing, deodorant, refreshing or antiseptic action. The medicated mouth washes should not be indiscriminately used by a normal person, the continuous use may prove harmful.

Enemas

Enemas are aqueous or oily solutions or suspensions intended for introduction into the rectum for their purgative, sedative, anthelmintic, anti-inflammatory or nutritive effects. They may also be used for X-ray examination of the lower bowel.

Inhalations

Inhalations are the liquid preparations containing volatile ingredients and are meant for local or systemic action on the nasal or respiratory tract. They are used to relieve nasal congestion and inflammation of the respiratory tract. If the ingredients are volatile at room temperature, they may be placed on an absorbent pad and inhaled therefrom. In other cases they may be added to warm water, but not boiling water and the vapours are inhaled for five to ten minutes.

Ear drops

Ear drops are the liquid preparations in which the drugs are dissolved or suspended in a suitable vehicle like water, dilute alcohol, glycerin, propylene glycol or any other suitable solvent and are intended for instillation into the ear with a dropper.

Ear drops are generally used for cleaning the ear, drying weeping surfaces, softening the wax and for treating the mild infections.

Ear drops are dispensed in coloured flutted bottles attached with a dropper or in suitable plastic containers. The containers should be labelled "for external use only".

Liniments

Liniments are the liquid preparations meant for external application to the skin. They are applied with friction and rubbing of the skin. They should not be applied to the broken skin. Liniments should be dispensed in coloured flutted bottles in order to distinguish from preparations meant for internal use. The bottle should be labelled "for external use only" and "shake the bottle before use".

Lotions

Lotions are usually liquid suspensions or dispersions intended for external application to the skin without rubbing. Lotions are used for local cooling, soothing or protective purposes.

Lotions should be dispensed in coloured flutted bottles in order to distinguish from preparations mean for internal use. The container should be labelled "for external use only". On long standing lotions have a tendency to separate out therefore the container must be labelled "shake before use".

Exercise No. 50

For :
Age :
Address :

℞

Orange tincture	6.0 ml
Simple syrup q.s.	100.0 ml

Make a syrup. Send 40 ml.

Type : Syrup.

Procedure

Mix orange tincture with simple syrup, stir so as to get a homogeneous preparation. Transfer to a bottle label and dispense.

Uses

It is used as flavouring/carminative syrup.

Exercise No. 51

For :
Age :
Address :

℞

Phenol glycerin	5 ml
Amaranth solution	1 ml
Water	to 100 ml

Label : The gargles.
Sig : Dilute it with an equal volume of warm water before use.
Type : Gargles.

Procedure

Mix Amaranth solution (1% w/v) with small amount of water, add phenol glycerin (16% w/w phenol and 84% w/w glycerin) and mix. To this incorporate more of vehicle to produce the required volume. Transfer to a container, label and dispense. Secondary label "for external use only" must be attached.

Uses

It is used as gargles for the treatment of pharynx and nasopharynx by forcing air from the lungs through the gargles which is held in the throat.

Phenol gargles should be diluted with an equal volume of warm water before use.

Exercise No. 52

For :
Age :
Address :

℞

Potassium chlorate	3.0 gm
Liquid phenol	1.5 ml
Water	to 100.0 ml

Label : The gargles. Send 50 ml.
Direction : One tablespoonful to be diluted with half a tumblerful of warm water before use.
Type : Gargles.

Procedure

Dissolve the weighed amount of potassium chlorate in about 15 ml of water. To this add liquefied phenol and add sufficient water to produce the required volume. Transfer to a container, label and dispense. The secondary label "for external use only" must be attached.

Uses

These gargles are used as sialogogue and astringent.

1. Potassium chlorate is a sialogogue (which increases the flow of saliva) and astringent (which precipitates the proteins).
2. These gargles should be diluted with ten times its volume of warm water before use.

Exercise No. 53

For :
Age :
Address :

℞

Sodium chloride	0.5 gm
Sodium bicarbonate	1 gm
Amaranth solution	2 ml
Peppermint water	to 100 ml

Label : The mouth wash.

Directions :

1. Dilute it with an equal volume of warm water before use.
2. Rinse the mouth 3-4 times daily as required.

Type : Mouth wash.

Procedure

Dissolve the weighed quantities of sodium chloride and sodium bicarbonate in 3/4th of the vehicle, add Amaranth solution and incorporate more of vehicle to produce the required volume. Transfer to a bottle, label and dispense. Attach "For external use only" label.

Uses

Compound sodium chloride mouth wash is a simple mouth wash which is used to clean and deodorise the buccal cavity. It is very refreshing particularly to the bed ridden patients.

1. Sodium chloride makes the preparation isotonic.
2. Sodium bicarbonate can dissolve mucous therefore added to spray solutions and washes for the throat and nose.
3. Amaranth solution is used as a colouring agent.
4. Peppermint water is used to impart pleasant taste and odour to the preparation.

Marketed preparation by :

Warner-Hindustan Division, Mumbai - 400072.

Listrine liquid

Contains :

Thymol	0.06% w/v
Eucalyptol	0.09% w/v
Methyl salicylate	0.06% w/v
Menthol	0.04% w/v
Benzoic acid	0.15% w/v
Alcohol	25.27% v/v

Exercise No. 54

For :
Age :
Address :

℞

Soft soap	25 gm
Purified water	to 500 ml

Label : The soap enema.
Sig : To be used as directed.
Type : Evacuant enema.

Procedure

Dissolve the soft soap in purified water. Transfer to a container, label and dispense. Attach the secondary label "for external use only".

Uses

It is used as an evacuant enema.

Precaution

Large volume enemas should be warmed to body temperature before administration.

Exercise No. 55

Prepare 500 ml of starch mucilage.

Formula

Starch	50 gm
Water	to 1000 ml

Label : The starch mucilage.
Type : Mucilage.

Procedure

Triturate the starch in a mortar with a small amount of water to form a smooth cream, add the rest of the vehicle and mix. Transfer this suspension to a flask and heat it to boiling with frequent agitation. The starch becomes gelatinised and forms a mucilage which should be free from lumps.

Remove the flask from heat and immediately cool it by rotating under tap water until cold; this prevents the formation of a 'skin'. Transfer to a measure and add more of warm water to produce the final volume. Pack in a container and label.

Uses

Starch mucilage is used as an emulsifying agent for oils and insoluble solids in enemas.

Exercise No. 56

For :
Age :
Address :

℞

Castor oil	20 ml
Starch mucilage	to 100 ml

Label : The enema. Send 300 ml.
Direction : To be used as directed.
Type : Castor oil enema.

Procedure

Triturate castor oil with a small amount of starch mucilage, add more of starch mucilage to produce the required volume. Triturate thoroughly so as to get a homogeneous product. Transfer to a bottle, label and dispense. Attach the secondary label "for external use only" and "shake the bottle before use".

Uses

It is used as purgative enema.

1. Castor oil soften the stools.
2. Starch mucilage acts as an emulsifying agent for the emulsification of castor oil, though it is not a true emulsifying agent. The use of starch mucilage as emulsifying agent is very limited, being usually restricted to enemas.

Exercise No. 57

For :
Age :
Address :

℞

Menthol	2 gm
Eucalyptus oil	10 gm
Light magnesium carbonate	7 gm
Water	to 100 ml

Make an inhalation. Send 50 ml.

Sig : One teaspoonful to be added to 500 ml hot water (about 65°C) and inhale the vapours.
Type : Aqueous inhalation.

Procedure

Dissolve the menthol in the eucalyptus oil, add and light magnesium carbonate and sufficient water to produce 100 ml. Stir well so as to get a homogeneous product. Transfer to a container, label with directions "For external use only" and "shake the bottle before use" and dispense.

Uses

It is used as an inhalation to remove congestion of the nostrils.

Light magnesium carbonate acts as a distributing agent to ensure uniform dispension of the oil on shaking. Light magnesium carbonate does not interfere with free volatilisation of the oil when the inhalation is added to hot water for use.

Exercise No. 58

For :
Age :
Address :

℞

Benzoin, crushed	10 gm
Prepared storax	5 gm
Alcohol (95 per cent)	to 100 ml

Label : The inhalation. Send 50 ml.
Direction : Put one teaspoonful to hot water (about 65°C) and inhale the vapours.
Type : Inhalation.

Procedure

Macerate the benzoin and the prepared storax with 75 ml of alcohol (95 per cent) for 24 hours. Filter, and pass sufficient alcohol through the filter to produce the required volume. Transfer to a container, attach "For external use only" label and dispense.

Uses

It is used as an inhalation, in the treatment of catarrh (cold etc.) of the upper respiratory tract.

Exercise No. 59

For :
Age :
Address :

℞

Phenol glycerin	4 ml
Glycerin to	10 ml

Label : The phenol ear drops. Send 20 ml.
Direction : Two drops to be instilled into affected ear as directed.
Type : Ear drops.

Procedure

Mix phenol glycerin with sufficient glycerin to make up the required volume. Pack in dropper bottle, label and dispense. Secondary label "For external use only" must be attached.

Uses

Phenol ear drops are used in the treatment of otitis media (an infection of the middle ear) and of boils in the ear.

Caution

Dilution with water renders phenol ear drops caustic. If need arises the preparation may be diluted with glycerin.

Formula for phenol glycerin

Phenol	16 gm
Glycerin	84 gm

Dissolve the phenol in glycerin with the help of gentle heat if necessary.

Exercise No. 60

For :
Age :
Address :

℞

Hydrogen peroxide solution (6 per cent)	12.5 ml
Purified water to	50.0 ml

Make ear drops. Send 10 ml.

Sig : One to two drops to be put into the left ear.
Type : Ear drops.

Procedure

Mix hydrogen peroxide solution with sufficient purified water to produce the required volume. Transfer to a droper bottle, label and dispense. Secondary label "for external use only" must be attached.

Uses

Hydrogen peroxide ear drops are used to remove wax from the ears.

Hydrogen peroxide solution acts as an antiseptic and deodorant. It is useful for detaching dead tissues and bacterial nests from dirty wounds. It is used as a deodorant gargle or mouth wash. For removing superfacial stains from the teeth and for bleaching hair, it should be mixed with an equal volume of water.

Marketed preparation by :

Arora Pharmaceuticals Pvt. Ltd., New Delhi - 110035.

Hydrogen peroxide solution
Each ml contains :

Conc. hydrogen peroxide 6% w/v.

Exercise No. 61

For :
Age :
Address :

℞

Chloramphenicol	5 gm
Propylene glycol to	100 ml

Label : The ear drops. Send 20 ml.
Sig : Three drops to be put into the affected ear as directed.
Type : Ear drops.

Procedure

Take the weighed quantity of chloramphenicol and dissolve in small amount of propylene glycol. Add more of propylene glycol to produce the required volume. Pack in a dropper bottle, label and dispense.

Uses

Chloramphenicol ear drops are used in chronic otorrhoea (blood discharge from the ear) and are often effective even when other antibacterial agents are ineffective.

Propylene glycol acts as a solvent.

Marketed preparation by :

Bombay Drug House Pvt. Ltd., Mumbai - 4000101

Each ml contains :

Chloramphenicol	5%.

Exercise No. 62

For :
Age :
Address :

℞

Boric acid	2 gm
Alcohol (95 per cent)	20 ml
Purified water to	100 ml

Label : The ear drops. Send 20 ml.
Signa : 3-4 drops to be put into ear occasionally.
Type : Ear drops.

Procedure

Dissolve the boric acid in about 20 ml of purified water, add alcohol and incorporate more of purified water to produce the required volume. Pack, label and dispense. Attach the secondary label "for external use only".

Uses

Boric acid has feeble anti-bacterial and antifungal properties. A solution in alcohol and water is used for instillation into ear for antibacterial and antifungal action.

Marketed preparation by :

Agrawal Pharmaceuticals, Delhi - 110092.

Boric acid ear drops.

Exercise No. 63

For :
Age :
Address :

℞

Ammonium chloride	1.25 gm
Dilute ammonia solution	4.50 ml
Oleic acid	8.50 ml
Turpentine oil	25.00 ml
Water	62.50 ml

Label : The white liniment.
Direction : To be applied externally on the affected part with friction.
Type : Emulsion type liniment made with an ammonium soap.

Procedure

Mix oleic acid and turpentine oil in a bottle, add dilute ammonia solution diluted with 4.5 ml warm water, and shake the bottle. Separately dissolve the ammonium chloride in remainder of the water and add to the emulsion. Mix thoroughly by shaking the bottle vigorously till a homogeneous emulsion is produced. Pack the bottle, label and dispense. Attach the secondary label "Shake the bottle before use" and "for external use only".

Uses

It is used as a rubefacient and counter-irritant for relief in rheumatic pains and stiffness.

1. In this prescription the soap is made by interaction of oleic acid and dilute ammonia solution during the preparation of liniment which is made by agitation of the bottle.
2. In the formula for white liniment the quantities produced are slightly more than 100 ml. Consider the quantities as such i.e. for 100 ml.

Exercise No. 64

For :
Age :
Address :

℞

Triethanolamine	1 gm
Oleic acid	4 gm
Benzyl benzoate	25 ml
Water to	100 ml

Make an application. Send 50 ml.

Direction : To be applied as directed.

Procedure

Triturate triethanolamine and oleic acid together in a mortar to form triethanolamine oleate (a soap). To this add benzyl benzoate, triturate until a thick white cream is obtained. To this incorporate 25 ml water, mix well. Add sufficient water to produce the required volume with continuous trituration so as to get a uniform product. Transfer to a bottle, cork, label and dispense. Apply the secondary label, "Shake the bottle before use" and "for external use only".

Uses

This application is used in the treatment of scabies. Scabies is a skin disease caused by a mite which burrows under the skin surface, causes extreme itching and discomfort all over the skin.

It is applied with a brush on whole of the body except face, after the body has been scrubbed with soft soap and hot water, to open up the burrows. The second coat is applied on the second day when the first has been dried up and left for 24 hours. After the second application the body is again thoroughly bathed with soap and hot water and clothes boiled in water. Although more than one course of treatment may be necessary, benzyl benzoate is more effective, cleaner and easier to apply than sulphur ointments.

Marketed preparation by :

Medo Chem Labs. Pvt. Ltd., Delhi - 110032.

1. Benzyl benzoate application
   Each contains :
   Benzyl benzoate 250 gm
   Emulsifying wax 20 gm
   Purified water to 1000 ml
2. Ascabiol emulsion (Rhone-Poulenc (India) Ltd., Mumbai-400025.
   Each contains :
   Benzyl benzoate 25%

Exercise No. 65

For :
Age :
Address :

℞

Calamine	15.0 gm	Zinc oxide	5.0 gm
Bentonite	3.0 gm	Sodium citrate	0.5 gm
Liquefied phenol	0.5 ml	Glycerin	5.0 ml
Purified water upto	100.0 ml		

Make : Lotion. Send 50 ml.
Direction : To be applied to the affected part of the skin without friction.

Procedure

Mix the weighed amount of calamine, zinc oxide and bentonite in a mortar. Triturate it with a solution of sodium citrate in about 70 ml of purified water. Add the required quantity of liquefied phenol and glycerin; mix well. To this add more of vehicle to produce the required volume, mix thoroughly so as to get a uniform preparation. Transfer to a bottle, cork, label and dispense. Apply the secondary label "Shake the bottle before use" and "for external use only".

Uses

This lotion is used as an astringent and protectant from sun burn. It acts as a smoothening agent and gives relief from itching and pain during skin irritation. It is also used in ringworm infection and eczema.

1. Calamine is basic zinc carbonate mixed with suitable amount of ferric oxide to impart pink colour. It is often prescribed by dermatologists to give flesh like colour to lotions or creams.
2. Zinc oxide has mild astringent, protective and antiseptic action. It is widely used in dusting powders, lotion, and ointments meant for the treatment of skin diseases and infections such as eczema, ring worm, psoriasis (chronic skin disease in wich red scaly patches develop) and pruritus (itching).
3. Bentonite is a native colloidal hydrated aluminium silicate. It is insoluble in water but swells up nearly seven times its bulk and forms a magma with desirable viscosity. Hence it is used as suspending agent for the dispersion of insoluble substances like calamine etc.
4. Sodium citrate is added to prevent the lotion from being too viscous. It acts as a buffer and maintains the pH appropriate for skin preparations.
5. Liquefied phenol acts as antipruritic because of its antiseptic properties and also because of its local anaesthetic action.
6. Glycerin acts as a humectant thus keeps the skin moist and have soothening effect on the skin.

Marketed preparation by :
Jilichem Labs (India) Ltd., Ahmedabad
Calamine lotion-formula is same as given above.

11

Pharmaceutical Incompatibilities

A pharmaceutical incompatibility may be defined as the result of prescribing or mixing the substances which are antagonistic in nature and an undesirable product is formed which may affect the safety, purpose or appearance of the preparation. These incompatibilities are of three types i.e. therapeutic, physical and chemical.

1. Therapeutic incompatibility

Therapeutic incompatibility may be the result of prescribing certain drugs to the patient with the intension to produce a specific degree of action but the nature or the intensity produced is different from that intended by the prescriber. It may be due to the administration of (i) over dose or improper dose of a single drug, (ii) wrong dose or dosage form, (iii) contra-indicated drugs, (iv) synergistic and antagonistic drugs. For example tetracycline hydrochloride administered with milk makes tetracycline inactive due to calcium present in milk. Therefore water should be used instead of milk for the oral administration of tetracycline hydrochloride preparations.

Sometimes haste or error in the case of prescribing a wrong drug e.g. mercuric chloride in place of mercurous chloride, barium carbonate in place of barium sulphate, labelling of skin ointment as eye ointment may be the cause of prescribing wrong drugs or mode of administration. Although for all these matters prescriber is responsible but it is the duty of the pharmacist to bring these errors to the notice of the prescriber for the instructions to be changed.

2. Physical incompatibility

Physical incompatibility is due to immiscibility, insolubility, precipitate formation or liquefication of solid materials. This usually causes nonuniform, unsightly or unpalatable mixture e.g. oils and water are immiscible with each other, which can be made miscible by the addition of an emulsifying agent. Indiffusible solids can be made diffusible by adding a suitable suspending agent. Eutectic substances can be dispensed by adding certain absorbent.

3. Chemical incompatibility

Chemical incompatibility may be the result of chemical reactions between the ingredients of a prescription and a harmful or even dangerous product may be formed. Therefore, precautions should be taken either to prevent the formation of harmful product or to correct them by addition or substitution of troublesome ingredient.

Chemical incompatibility may be :

(a) Tolerated

In which the reaction is minimised by adopting some suitable order of mixing or mixing the solutions in dilute form but no alteration is made in the active ingredients of the preparation.

(b) Adjusted incompatibility

In such incompatibilities the reaction is prevented by addition or substitution of one of the reacting substances with another of equal therapeutic value but does not affect the medicinal action of the preparation (e.g. substitution of caffeine citrate with caffeine in sodium salicylate and caffeine citrate mixture).

In some of the chemical reactions precipitates are formed which may be diffusible or indiffusible in nature.

If the precipitates formed are diffusible in nature then method A is followed in which the vehicle is divided in two parts, to one part one of the reacting substances is dissolved and to the other part other reacting substances are dissolved and then these two solutions are mixed together slowly with continuous stirring.

If the precipitates formed are indiffusible in nature then method B is followed in which the vehicle is divided into two parts, to one part one of the reacting substance is dissolved and to the other part a suspending agent like compound tragacanth powder in the ratio of 2 gm/100 ml or tragacanth mucilage in the ratio of 1/4th of the total volume of mixture to be prepared is triturated with troublesome ingredient, then the first and second solutions are combined together slowly with continuous stirring.

Whether method A or method B has been used in dispensing the prescription, it is very important to fix a "Shake the bottle" label to the container and ensure that the patient follows and realizes the importance of these directions.

Exercise No. 66

For :
Age :
Address :

℞

Castor oil	30 ml
Water to	120 ml

Make an emulsion. Send 60 ml

Sig : Cochleare amplum omni nocte sumenda.
Type : Physical incompatibility.

Explanation

In this prescription castor oil is a fixed oil which is immiscible with water, to make them miscible an emulsifying agent will have to be added, for which gum acacia is the most suitable emulsifying agent.

Procedure

Triturate the calculated quantity of gum acacia with calculated quantity of water required for primary emulsion to form a mucilage. To this add castor oil in small quantities at a time with constant trituration till a clicking sound is produced. At this stage the emulsion is known as primary emulsion. Dilute the primary emulsion by adding more of vehicle. Transfer the emulsion to a measure and incorporate more of vehicle to produce the required volume. Transfer the emulsion to a bottle, cork, label and dispense Attach the secondary label, "Shake the bottle before use".

Uses

Castor oil is used as a purgative.

Exercise No. 67

For :
Age :
Address :

℞

Menthol	5 gm
Camphor	5 gm
Ammonium chloride	30 gm
Light magnesium carbonate	60 gm

Make an insufflation. Send 25 gm

Sig : More dicto sumenda.
Type : Physical incompatibility.

Explanation

In the above prescription menthol, camphor and ammonium chloride are liquefiable substances (eutectic substances). When any two of them are combined together a eutectic mixture is formed due to the reason that the new compound so formed has lower melting point than any of the mixed substances. Therefore to dispense them in powders, light magnesium carbonate is added which acts as an absorbent.

Procedure

Separately powder menthol, camphor and ammonium chloride. Add one by one these powdered substances to light magnesium carbonate with thorough trituration. Mix thoroughly until uniform. Transfer to a container, label and dispense.

Uses

It is used as an insufflation.

Exercise No. 68

For :
Age :
Address :

℞

Codeine phosphate 0.5 gm

Flat pulvis. Mitte tales sexies.

Sig : Unus omni nocte sumenda.
Type : Therapeutic incompatibility.

Explanation

In this prescription overdose of codeine phosphate has been prescribed. Probably the physician intended to write 5 milligrammes but out of mind he wrote 0.5 gm instead of 0.005 gm of codeine phosphate. Therefore this prescription should not be dispensed and must be referred back to the prescriber.

Exercise No. 69

For :
Age :
Address :

℞

Strychnine hydrochloride solution	6 ml
Aromatic spirit of ammonia	4 ml
Water ad	120 ml

Make a mixture. Send 60 ml.

Sig : Cochleare amplum bis in die sumenda.
Type : Chemical incompatibility (incompatibility of alkaloidal salts with alkaline substances).

Explanation

Strychnine hydrochloride is an alkaloidal salt and aromatic spirit of ammonia is an alkaline substance. When they react together the precipitates of strychnine are formed because the quantity of strychnine hydrochloride prescribed is much more than its solubility which is 1 in 60, moreover the amount of alcohol present in aromatic spirit of ammonia is also negligible, hence strychnine gets precipitated. The precipitates so formed are diffusible in nature. Therefore follow method A for precipitate yielding combinations.

Procedure

Divide the vehicle in two equal parts, to one part dissolve strychnine hydrochloride solution and to the other part dissolve aromatic spirit of ammonia. Combine the two solutions with continuous stirring. Transfer the mixture to the bottle, label and dispense. Attach "Shake the bottle before use" label.

Uses

Strychnine hydrochloride acts as a bitter tonic and aromatic spirit of ammonia acts as a stimulant.

Exercise No. 70

For :
Age :
Address :

℞

Quinine hydrochloride	130 mg
Sodium salicylate	4 gm
Water upto	90 ml

Make a mixture. Send 60 ml.

Sig : Cochleare maximum t.d.s.
Type : Chemical incompatibility (incompatibility of alkaloidal salts with salicylates).

Explanation

Quinine hydrochloride reacts with sodium salicylate to form the precipitates of quinine salicylate which are indiffusible in nature. Therefore, follow method B for precipitate yielding combinations.

Procedure

Divide the vehicle in two parts, to one part dissolve quinine hydrochloride and to the other part triturate calculated amount of compound tragacanth powder and sodium salicylate. Combine the two portions slowly with thorough trituration. Transfer the mixture of a bottle, label and dispense. Attach the secondary label "Shake the bottle before use".

Uses

This mixture is used as an antimalarial mixture. Sodium salicylate acts as an analgesic, antipyretic and antirheumatic.

Exercise No. 71

For :
Age :
Address :

℞

Sodium salicylate	1.0 gm
Caffeine citrate	0.650 gm
Water upto	30.0 ml

Make a draught.

Sig : Statim sumat.
Type : Chemical (adjusted) incompatibility.

Explanation

Caffeine citrate is a mixture of equal weights of caffeine and citric acid. The citric acid present reacts with sodium salicylate to liberate salicylic acid which gets precipitated. But if caffeine is used instead of caffeine citrate, it forms a soluble complex with sodium salicylate. Therefore, substitute half as much caffeine as that of caffeine citrate and a clear mixture will be obtained.

Procedure

Dissolve sodium salicylate in small amount of water, to this add caffeine with continuous stirring. Add more of vehicle to produce the required volume. Transfer the mixture to a bottle, label and dispense.

Uses

It is used as an analgesic and antirheumatic mixture.

Exercise No. 72

For :
Age :
Address :

℞

Ferric chloride solution	1 ml
Potassium iodide	2 gm
Potassium citrate	4 gm
Water to produce	60 ml

Fiat : Mistura. Mitte 90 ml.
Signa : 15 ml pro dose bis in die sumenda.
Type : Incompatibility of iodides with ferric salts.

Explanation

When ferric chloride solution is combined with potassium iodide, free iodide is liberated which gets precipitated but if ferric chloride is converted to an organic compound before the addition of potassium iodide then no precipitation takes place. Therefore mix potassium citrate solution with ferric chloride solution and then add potassium iodide solution. The resulting solution will be green due to the formation of a new compound by ferric chloride and potassium citrate.

Procedure

Dissolve potassium citrate in about 3/4th of the vehicle, to this add ferric chloride solution and then add potassium iodide with continuous stirring. Add more of vehicle to produce the final volume. Transfer the mixture to a bottle, label and dispense. Attach "shake the bottle before use" label.

Uses

This mixture is used in goiter (an enlargement of the thyroid gland).

Exercise No. 73

For :
Age :
Address :

℞

Sodium salicylate	2 gm
Sodium bicarbonate	4 gm
Ferric chloride solution	0.5 ml
Water to	90 ml

Fiat : Mistura. Mitte 60 ml.
Sig : More dicto sumenda.
Type : Incompatibility of soluble salicilates with ferric salts.

Explanation

Ferric chloride reacts with sodium salicylate to form precipitates of ferric salicylate which are indiffusible in nature but the precipitates so formed become soluble in the presence of sodium bicarbonate, hence a clear mixture is formed.

Procedure

Dissolve sodium salicylate and sodium bicarbonate in 3/4th of the vehicle, add ferric chloride solution and notice considerable effervescence. When effervescence ceases, filter through cotton wool and pass more of vehicle through the filter to produce the required volume. Transfer the solution to a bottle, label and dispense. Attach "Shake the bottle before use" label.

Uses

It is used as an antirheumatic mixture.

Exercise No. 74

For :
Age :
Address :

℞

Sodium bicarbonate	1.5 gm
Borax	1.5 gm
Phenol	0.75 gm
Glycerin	25.00 gm
Water upto	100.00 ml
Make a spray solution. Send	50 ml

Type : Incompatibility of evolution of carbon dioxide.

Explanation

When sodium bicarbonate, borax and glycerin are mixed together in the presence of water, a reaction takes place with the evolution of carbon dioxide. If the mixture is dispensed as such, there are chances of bursting the bottle. Therefore, mix these substances in an open vessel until evolution of carbon dioxide ceases.

Procedure

Dissolve sodium bicarbonate and borax in sufficient amount of water in an open vessel, add glycerin and stir, wait until evolution of carbon dioxide ceases. Incorporate phenol, mix and transfer the solution to a bottle, label and dispense.

Uses

It is used as an antiseptic spray solution.

Exercise No. 75

For :
Age :
Address :

℞

Phenobarbitone sodium	650 mg
Ammonium bromide	8 gm
Water upto	120 ml

Make a mixture. Send 90 ml.

Sig : Cochleare amplum hora somni sumenda.

Type : Incompatibility of soluble barbiturates with ammonium bromide.

Explanation

In this prescription phenobarbitone sodium reacts with ammonium bromide with the formation of indiffusible precipitates of phenobarbitone, but if ammonium bromide is replaced with an equivalent amount of sodium bromide or potassium bromide, a clear mixture is obtained. The pharmacological action of these three bromides is same but the intensity of action depends on the amount of bromide radicle used.

$$NH_4Br = NaBr$$

$$98 = 103 \qquad \text{(Mol wt. of } NH_4Br = 98\text{; Mol wt. of } NaBr = 103\text{)}$$

For 98 gm of NH_4Br the amount of NaBr required = 130 gm

For 1 gm of NH_4Br the amount of NaBr required $= \frac{103}{98}$ gm

For 8 gm of NH_4Br the amount of NaBr required $= \frac{103}{98} \times 8$ gm

$= 8.4$ gm

Therefore, replace 8 gm of ammonium bromide with 8.4 gm of sodium bromide and dispense the mixture as such. The resulting mixture will be a clear mixture.

Procedure

Dissolve phenobarbitone sodium in 3/4th of the vehicle, add sodium bromide, stir until dissolved. Incorporate more of vehicle to produce the required volume. Transfer to a bottle, label and dispense.

Uses

This mixture is used as hypnotic and sedative.

12

Suppositories

Suppositories are special shaped solid dosage form of medicament for insertion into body cavities other than mouth. They may be inserted into rectum, vagina or urethra. These products are so formulated that after insertion, they will either melt or dissolve in the cavity fluids to release the medicament. Suppositories varies in shapes, sizes and weights. Generally suppositories weighing 1 to 2 gm are prepared. The bases used for the preparation of suppositories include cocoa butter, glycero gelatin and soap glycerin.

Cocoa butter also known as theobrome oil is obtained from the crushed and roasted seeds of Theobroma cocoa. It has butter like consistency and chocolate like odour. It has a melting point of 30 to 35°C.

Uses

Suppositories are used for any one of the three different purposes.

1. To produce local action.
2. To produce systemic action.
3. To produce mechanical action on the lower bowel and facilitate evacuation in the treatment of haemorrhoids, anal irritation, constipation etc.

Types of suppositories

1. Rectal suppositories

They are meant for introduction into the rectum for their systemic action. They are tapered at one or both ends and usually weigh about 2 gm. Rectal suppositories meant for children are smaller in size and weight than the adult suppositories. They usually weigh about 1 gm.

2. Vaginal suppositories

They are also known as pessaries and are meant for introduction into the vagina. Now a days a few special tablets or capsules, oval or suppository shaped are prepared for use in the vagina and are known as vaginal tablets or vaginal capsules respectively.

3. Urethral suppositories

They are also known as urethral bougies and are meant for introduction in the urethra. They are very rarely used.

4. Nasal suppositories

They are also known as nasal bougies or buginaria and are meant for introduction into the nasal cavity. They are always prepared with glycero-gelatin base. Nasal suppositories are very rarely used.

5. Ear cones

They are also known as aurinaria and are meant for introduction into the ear. They are seldom used.

Suppository bases

Since suppositories are special solid dosage form of medicament they must retain its shape, solidity and firmness during storage and administration but must melt or dissolve in the cavity fluids when inserted into the body cavity. Therefore the materials used as suppository bases must impart these properties and also fulfil other formulation requirements. There are a large number of bases used but theobroma oil, glycerogelation and polyethylene glycols fulfils the above mentioned requirements.

1. Theobroma oil

It is also known as cocoa butter. It is obtained from the crushed and roasted seeds of Theobroma cocoa. It is a yellowish white solid which becomes brittle on storage. It has butter like consistency and chocolate like odour. It has a melting point of 30 to 35°C. It is a mixture of the glyceryl esters of stearic, palmitic, oleic and other fatty acids.

2. Glycerogelatin

Glycerogelatin base is a mixture of glycerin and water which is made stiff by the addition of gelatin. The stiffness of the mass depends upon the proportion of gelatin used. Glycerogelatin base is well suited for suppositories containing belladonna extract, boric acid, chloral hydrate, bromides, iodides, iodoform etc. This base may be used to prepare all types of suppositories but it is particularly used in vaginal suppositories.

3. Soap-glycerin suppositories

In glycero-gelatin base the gelatin is replaced with either curd soap or sodium stearate which makes the glycerin sufficiently hard for suppositories and a large quantity of glycerin upto 95% of the mass can be incorporated, further the soap helps in the evacuation action of glycerin whereas the gelatin does not. Soap-glycerin suppositories are very hygroscopic therefore must be wrapped in waxed paper on tin foil.

4. Emulsifying bases

They are synthetic bases and a number of proprietary bases of good quality are available e.g. massa esterinum, witepsol and massuppol.

Preparation of suppositories

Suppositories are prepared by hot process and cold compression. On laboratory scale they are prepared by hot process or fusion method. In this method a suppository mould made up of stainless steel, nickel-copper alloy, brass, aluminium, or plastic having 6 to 12 cavities with desired shape and size is used. Unless otherwise stated a standard mould of 1 gm capacity is used.

For cleaning, lubrication and removal of suppositories the mould can be opened longitudinally by removing the screws in the centre of the plates. For cleaning the opened plates are cleaned with detergent and hot water, and dried thoroughly. During cleaning the mould should never be scrapped.

The thoroughly cleaned and dried mould is then lubricated with opposite type of lubricant than base e.g. a lubricant containing soft soap 10 gm, glycerol 10 gm and alcohol (90%) 50 ml is most suitable for oily bases and liquid paraffin or arachis oil for glycero-gelatin suppositories respectively.

Before melting the base, keep the lubricated mould in the inverted position over ice to cool and drain and excess of the lubricant. Taking into consideration the displacement value of the medicament, place the calculated quantities of the base in a dish heated over water bath. To this add the weighed amount of drug and mix to melt. Pour the melted mass into the cavities of the mould already kept on ice. Fill the required number of cavities to overflowing, this is done to prevent the formation of hollows in the top of finished suppositories because cocoa butter contracts on cooling due to which hollows are formed. Allow to remain the mould on ice for about 15 minutes or till the mass solidifies. Remove the mould from ice and trim off excess of the mass with the help of a sharp blade or knife. Open the mould and remove the suppositories. If any lubricant is there, wipe it off with filter paper or clean cloth. Wrap the suppositories individually in wax paper and then pack them in shallow, partitioned card board boxes.

Displacement value

Since the volume of a suppository from a particular mould remains same but its weigh varies due to the variation in densities of medicaments and the base with which the mould was calibrated. To get a product of uniform and accurate weight, allowance must be made for the change in density of the mass due to added drugs. For this purpose the displacement value of the medicament is taken into consideration.

Displacement value may be defined as the quantity of the drug which displaces one part of the medicament.

Displacement values of some medicaments used in suppositories with reference to cocoa butter are given below :

Medicament	*Displacement value*
Aminophylline	1.5
Boric acid	1.5
Chloral hydrate	1.5
Cocaine hydrochloride	1.5
Hamamelis dry extract	1.5
Hydrocortisone acetate	1.5
Iodoform	4.0
Morphine hydrochloride	1.5
Phenobarbitone	1.0
Tannic acid	1.0
Zinc oxide	5.0
Castor oil	1.0
Liquid medicaments	1.0

To calculate the actual quantity of base required the following procedure is used.

Prepare 8 suppositories each containing 300 mg of bismuth subgallate. The displacement value of bismuth subgallate is 3.

To take into consideration wastage, calculate for two extra suppositories i.e. for 10.

Since 1 gm weight suppository mould is used so

Total weight of theobroma oil (alone) required for 10 suppositories	$= 1 \times 10 = 10$ gm
Total weight of bismuth subgallate required for 10 suppositories	$= 300 \times 10$
	$= 3000$ mg $= 3$ gm
Displacement value of bismuth subgallate	$= 3$
i.e. 3 gm of bismuth subgallate displaces	$= 1$ gm of cocoa butter
1 gm of bismuth subgallate displaces	$= 1/3$ gm of cocoa butter
3000 mg or 3 gm of bismuth subgallate displaces	$= 1/3 \times 3$
	$= 1$ gm of cocoa butter
Using 1 gm mould actual amount of cocoa butter required	$= (1 \times 10) - 1 = 9$ gm

The total weight of 10 suppositories will be $9 + 3 = 12$ gm. Therefore although made in a 1 gm mould each suppository weighs 1.2 gm. Using standard mould it is impracticable to make suppositories each containing 300 mg of the drug and weighing 1 gm. Therefore in prescriptions 1 gm indicates the standard size of the mould and necessary calculations according to displacement value of the medicament must be made for correct quantity of base to be used in the preparation of suppositories.

Exercise No. 76

For :
Age :
Address :

℞

Boric acid	200 mg
Theobroma oil	q.s.

Prepare suppositories. Send such five.

Sig : Unus more dicto sumenda.

Displacement value of boric acid is 1.5.

Procedure

To take into consideration the wastage calculate for six suppositories instead of five.

Melt the calculated quantity of theobroma oil in a dish over water bath. To this add boric acid and mix thoroughly. Pour this melted mass into the cavities of the mould already lubricated and kept on ice. Fill six suppositories' cavities to overflowing. Allow the mass to solidify. After about 5-10 minutes when the mass has solidified, trim off excess of the mass with the help of a sharp blade or knife. Open the mould and remove the suppositories. Wipe if any lubricant is there with filter paper or clean cloth. Wrap the suppositories individually in wax paper and then pack them in partitioned card board boxes.

Uses

These are used as antibacterial and antifungal suppositories.

Exercise No. 77

For :
Age :
Address :

℞

Alum	300 mg
Theobroma oil	q.s.

Fiat : Suppositorium. Mitte tales quarta.
Sig : Unus omni nocte utendum.

Displacement value of alum is 2.0.

Procedure

To take into consideration the wastage calculate for five suppositories instead of four.

Melt the calculated quantity of theobroma oil in a dish over water bath. Pour about half of the melted theobroma oil on powdered alum already placed on a tile, mix thoroughly. Transfer the mixed mass to the dish, if necessary warm to make the mass pourable. Pour the melted mass into the cavities of the mould already lubricated and kept on ice. Fill five cavities to overflowing. Allow the mass to solidify. When the mass has solidified, trim off excess of the mass with a sharp blade or knife. Open the mould and remove the suppositories. If any lubricant is present wipe it off with filter paper or clean cloth. Wrap the suppositories individually in wax paper and then pack in partitioned card board boxes.

Uses

Alum suppositories are used as an astringent.

Exercise No. 78

For :
Age :
Address :

℞

Chloral hydrate	200 mg
Theobroma oil	q.s.

Fiat : Suppositorium. Mitte tales sexies.
Signa : More dicto utenda.

Displacement value of chloral hydrate is 1.5.

Procedure

To take into consideration the wastage calculate for eight suppositories instead of six.

Melt the calculated amount of theobroma oil in a dish over water bath. Pour about half of the melted theobroma oil on chloral hydrate already placed on a tile, mix thoroughly. Transfer the mixed mass to the dish, if necessary warm to make the mass pourable. Pour the melted mass into the cavities of the mould already lubricated and kept on ice. Fill five cavities to overflowing. Allow the mass to solidify. When the mass has solidified, trim off excess of the mass with a sharp blade or knife. Open the mould, remove the suppositories. If any lubricant is present, wipe it off with filter paper or clean cloth. Wrap the suppositories individually in wax paper and then pack in partitioned card board boxes.

Uses

These suppositories are used as hypnotic.

Exercise No. 79

For :
Age :
Address :

℞

Gelatin	1.4 gm
Glycerin	7.0 gm
Purified water	a sufficient quantity.

Mix and prepare suppositories. Divide into ten suppositories.

Direction : One suppository to be inserted into the rectum at bed time.

Calculating the quantities

Glycerogelatin base is denser than theobroma oil and a 1 gm mould will hold 1.2 gm of glycerin suppositories mass. This fact must be taken into account in calculating the required weight of bases. When appropriate, displacement values are used in the usual way and after the amount of the base has been calculated, it is multiplied by the density factor 1.2 to get the required quantities of the base to be used.

Procedure

Soak the gelatin in about 3 ml purified water in a tared dish for about 10 minutes. Drain off the surplus water and add glycerin previously heated to 100°. Heat it on water bath until gelatin dissolves, evaporate the contents of the dish till required weight is obtained, if necessary add hot purified water to get required weight. Avoid excess stirring which may introduce air bubbles into the finished suppositories.

Pour the melted mass into the cavities of the mould which has been previously lubricated with liquid paraffin or arachis oil and kept in the inverted position on ice. Fill the required number of cavities. Overflowing the cavities is unnecessary because contraction on cooling is inappreciable so no hollows are formed rather it is difficult to remove excess mass neatly by cutting. Allow the mass to solidify. When the mass has solidified within 10-15 minutes, open the mould and remove the suppositories. Wrap them individually in tin foil, label and dispense.

Uses

These suppositories are used to evacuate the bowel.

Exercise No. 80

For :
Age :
Address :

℞

Tannic acid	0.2 gm
Theobroma oil	Sufficient quantity.

Fiat suppositorium. Mitte tales octo.

Displacement value of tannic acid is 1.0.

Procedure

To take into consideration the wastage calculate for ten suppositories instead of eight.

Melt the calculated amount of theobroma oil in a dish over water bath. Keep the weighed amount of tannic acid on an ointment slab. Pour half the melted theobroma oil over the tannic acid, mix thoroughly with a spatula. After thorough mixing, put the melted mass in the dish already containing remaining half of the base. If necessary warm to make the mass pourable. Pour the mixed mass into ten cavities of the lubricated mould to overflowing. Allow the mass to solidify. When the mass has solidified, trim off excess of the mass with a sharp blade or knife. Open the mould, remove the suppositories. If any lubricant is present over the suppositories, wipe it off with filter paper or clean cloth. Wrap the suppositories individually in wax paper and then pack in partitioned card board boxes. Label and dispense.

Storage

Preserve tannic acid glycerin suppositories in airtight containers.

Uses

Tannic acid suppositories are used as astringent.

Exercise No. 81

For :
Age :
Address :

℞

White bees wax	1.8 gm
Borax	0.1 gm
Liquid paraffin	6.1 gm
Rose oil	0.01 ml
Rose water	2.0 ml

Prepare cold cream. Send 10 gm.

Theory

Borax soaps are generally made by the interaction of borax and the free acid present in beeswax. Sodium soaps usually produce o/w emulsion but in this case phase inversion accurs and w/o cold cream is obtained.

Cold cream of borax beeswax is not very stable and cracks when rubbed on skin. The aqueous phase slowly evaporates producing a cooling effect. Therefore called cold cream. The residual oily film has an emollient action.

White beeswax and borax act as emulsifying agents.

Rose water is a perfume and acts as aromatic agent.

Uses

It acts as an emollient.

Exercise No. 82

For :
Age :
Address :

℞

Stearic acid	15 gm
Mineral oil	1 gm
Span 80	2 gm
Tween 60	1.5 gm
Propyl paraben	0.1 gm
Methyl paraben	0.2 gm
Sorbitol solution (70%)	3 gm
Perfume	q.s. (1-2 drops)
Water	100 gm

Prepare and supply 25 gm of vanishing cream.

Theory

Vanishing creams are o/w emulsions. On application water evaporates from the skin surface thereby leaving no trace of application. They are mainly used in summers.

Stearic acid acts as an emollient. Span 80 (oil soluble) and tween 60 (water soluble) act as emulsifying agents.

Methyl paraben and propyl paraben act as preservatives and are usually combined in the ratio of 1 : 9 (methyl paraben : propyl paraben). Sorbitol acts as humectant.

Exercise No. 83

For :
Age :
Address :

℞

Acetone	75%
Ethyl acetate	16%
Propylene glycol	4%
Water	5%

Prepare and supply nail lacquer remover 25 ml.

Theory

Acetone acts as a solvent for the nitrocellulose present in nail polish.

Ethyl acetate has soothing and cooling effect and also acts as solvent for nitrocellulose.

Propylene glycol has been added to prevent dehydration by acetone.

These may be liquid or creams. The liquid is mixture of solutions of varying volatilities. Liquid remover usually contain higher proportion of acetone, butyl or alkyl acetone. They may be containing vegetable oil or fatty esters as emollients.

Uses

It is used to remove nail lacquer.

Exercise No. 84

For :
Age :
Address :

℞

A. Oil soluble	Stearic acid	13.8 gm
	Stearyl alcohol	2.0 gm
	Isopropyl myristate	1.9 gm
	Mineral oil	2.0 ml
	Lanolin	2.0 gm
	Span 60	1.0 gm
B. Water soluble	Tween 60	2.4 gm
	Triethanolamine	1.0 gm
	Sodium lauryl sulphate	1.0 gm
	Water	72.9 ml
	Methyl paraben	q.s.
C.	Perfume	q.s.

Prepare 25 gm of beard softner (pre-shave product).

Theory

- **Stearic acid :** It acts as an emollient.
- **Stearyl alcohol :** It is surface active agent used to stabilize emulsions and has ability to retain large amount of water.
- **Isopropyl myristate :** It acts as pharmaceutical aid, used in cosmetics and topical medicinal preparations as an emollient, lubricant and enhances absorption through skin.
- **Lanolin :** It forms o/w emulsion and emulsifies aqueous liquids.
- **Span 60 :** Sorbitan esters, non-ionic surfactant and used as emulsifying agent in the preparation of w/o emulsion.
- **Tween 60 :** Sorbitan esters, polyoxyethylene derivatives. Non-ionic surfactant, hydrophyllic emulsifying agent forming o/w emulsion.
- **Triethanolamine :** In combination with fatty acids e.g. oleic acid as an emulsifier. A solvent for fats, oils, with fatty acids forms soaps and finds application in various types of emulsions, lotions, creams etc.
- **Sodium auryl sulphate :** It acts as surfactant and wetting agent.
- **Methyl paraben :** It acts as preservative.

Exercise No. 85

For :
Age :
Address :

℞

Ethanol denatured	50%
Sorbitol (70%)	2.5%
Perfume	0.5%
Methanol	0.1%
Boric acid	2.0%
Demineralised water	44.9%

Prepare 25 ml after-shave lotion.

Theory

- **Sorbitol 70% :** It acts as vehicle, plasticizer, humectant.
- **Boric acid :** It acts as weak germicide.
- **Methanol :** It is a vehicle.
- **Perfume :** Flavouring agent.
- **Denatured alcohol :** It acts as solvent for perfume.